Information Assurance for NETWORK-CENTRIC NAVAL FORCES

Committee on Information Assurance for Network-Centric Naval Forces

Naval Studies Board

Division on Engineering and Physical Sciences

NATIONAL RESEARCH COUNCIL
OF THE NATIONAL ACADEMIES

THE NATIONAL ACADEMIES PRESS
Washington, D.C.
www.nap.edu

THE NATIONAL ACADEMIES PRESS 500 Fifth Street, N.W. Washington, DC 20001

NOTICE: The project that is the subject of this report was approved by the Governing Board of the National Research Council, whose members are drawn from the councils of the National Academy of Sciences, the National Academy of Engineering, and the Institute of Medicine. The members of the committee responsible for the report were chosen for their special competences and with regard for appropriate balance.

This study was supported by Contract No. N00014-05-G-0288, DO #19 between the National Academy of Sciences and the Department of the Navy. Any opinions, findings, conclusions, or recommendations expressed in this publication are those of the author(s) and do not necessarily reflect the views of the organizations or agencies that provided support for the project.

International Standard Book Number-13: 978-0-309-13663-1
International Standard Book Number-10: 0-309-13663-6

Copies of this report are available from:

Naval Studies Board, National Research Council, The Keck Center of the National Academies, 500 Fifth Street, N.W., Room WS904, Washington, DC 20001; and

The National Academies Press, 500 Fifth Street, N.W., Lockbox 285, Washington, DC 20055; (800) 624-6242 or (202) 334-3313 (in the Washington metropolitan area); Internet, http://www.nap.edu.

Printed in the United States of America

THE NATIONAL ACADEMIES

Advisers to the Nation on Science, Engineering, and Medicine

The **National Academy of Sciences** is a private, nonprofit, self-perpetuating society of distinguished scholars engaged in scientific and engineering research, dedicated to the furtherance of science and technology and to their use for the general welfare. Upon the authority of the charter granted to it by the Congress in 1863, the Academy has a mandate that requires it to advise the federal government on scientific and technical matters. Dr. Ralph J. Cicerone is president of the National Academy of Sciences.

The **National Academy of Engineering** was established in 1964, under the charter of the National Academy of Sciences, as a parallel organization of outstanding engineers. It is autonomous in its administration and in the selection of its members, sharing with the National Academy of Sciences the responsibility for advising the federal government. The National Academy of Engineering also sponsors engineering programs aimed at meeting national needs, encourages education and research, and recognizes the superior achievements of engineers. Dr. Charles M. Vest is president of the National Academy of Engineering.

The **Institute of Medicine** was established in 1970 by the National Academy of Sciences to secure the services of eminent members of appropriate professions in the examination of policy matters pertaining to the health of the public. The Institute acts under the responsibility given to the National Academy of Sciences by its congressional charter to be an adviser to the federal government and, upon its own initiative, to identify issues of medical care, research, and education. Dr. Harvey V. Fineberg is president of the Institute of Medicine.

The **National Research Council** was organized by the National Academy of Sciences in 1916 to associate the broad community of science and technology with the Academy's purposes of furthering knowledge and advising the federal government. Functioning in accordance with general policies determined by the Academy, the Council has become the principal operating agency of both the National Academy of Sciences and the National Academy of Engineering in providing services to the government, the public, and the scientific and engineering communities. The Council is administered jointly by both Academies and the Institute of Medicine. Dr. Ralph J. Cicerone and Dr. Charles M. Vest are chair and vice chair, respectively, of the National Research Council.

www.national-academies.org

COMMITTEE ON INFORMATION ASSURANCE FOR NETWORK-CENTRIC NAVAL FORCES

BARRY M. HOROWITZ, University of Virginia, *Co-Chair*
NILS R. SANDELL, JR., BAE Systems Advanced Information Technologies, *Co-Chair*
M. BRIAN BLAKE, Georgetown University
CLYDE G. CHITTISTER, Carnegie Mellon University, Software Engineering Institute
ANUP K. GHOSH, George Mason University
RAYMOND HALLER, MITRE
RICHARD J. IVANETICH, Institute for Defense Analyses
JOHN W. LINDQUIST, EWA Information and Infrastructure Technologies, Inc.
MARK W. MAIER, The Aerospace Corporation
RICHARD W. MAYO, USN (retired), CACI International, Inc.
ANN K. MILLER, Missouri University of Science and Technology
DANIEL M. SCHUTZER, Financial Services Technology Consortium
RALPH D. SEMMEL, Johns Hopkins University Applied Physics Laboratory
ROBERT M. SHEA, USMC (retired), Smartronix
JOHN P. STENBIT, Independent Consultant, Oakton, Virginia
SALVATORE J. STOLFO, Columbia University
EDWARD B. TALBOT, Sandia National Laboratories
DAVID A. WHELAN, The Boeing Company

Staff

CHARLES F. DRAPER, Director, Naval Studies Board
BILLY M. WILLIAMS, Study Director
RAYMOND S. WIDMAYER, Senior Program Officer
SUSAN G. CAMPBELL, Administrative Coordinator
MARY G. GORDON, Information Officer
SEKOU O. JACKSON, Senior Project Assistant
SIDNEY G. REED, JR., Consultant

Marine Corps Liaison Representative

LTGEN JAMES F. AMOS, USMC, Commanding General, Marine Corps Combat Development Command (through July 2, 2008)
LTGEN GEORGE J. FLYNN, USMC, Commanding General, Marine Corps Combat Development Command (as of July 28, 2008)

Staff

CHARLES F. DRAPER, Director
RAYMOND S. WIDMAYER, Senior Program Officer
BILLY M. WILLIAMS, Senior Program Officer
MARTA V. HERNANDEZ, Associate Program Officer
SUSAN G. CAMPBELL, Administrative Coordinator
MARY G. GORDON, Information Officer
SEKOU O. JACKSON, Senior Program Assistant

Preface

Long before naval leaders began articulating network-centric warfare as a concept,[1] the U.S. Navy integrated weapons and sensors at diverse locations to perform its missions. For example, in the mid-20th century, antisubmarine warfare operations depended on long-range but limited-accuracy sensors cueing an air platform so that it could deploy shorter-range but more-accurate sensors capable of yielding improved targeting. Today's accelerating pace of advances in computing and communications capabilities has led to an even broader vision of network-centric operations that includes all military force operations in peace as well as war and in which network centric operations have been defined as "military operations that exploit state-of-the-art information and networking technology to integrate widely dispersed human decision makers, situational and targeting sensors, and forces and weapons into a highly adaptive, comprehensive system to achieve unprecedented mission effectiveness."[2]

One of the key attributes of network-centric operations is a reliable and robust capability to support well-informed and rapid decision making by military commanders at all levels, within a system of flexible and adaptable command relationships. Underlying this attribute, of course, is the need to ensure that accurate information is securely gathered, distributed, and stored in ways that are timely, trustworthy, and not subject to disruption, corruption, or exploitation by

[1]For example, see VADM Arthur K. Cebrowski, USN; and John J. Garstka, 1998, "Network-Centric Warfare: Its Origin and Future," *U.S. Naval Institute Proceedings,* January, pp. 28-35.

[2]Naval Studies Board, National Research Council. 2000. *Network-Centric Naval Forces: A Transition Strategy for Enhancing Operational Capabilities,* National Academy Press, Washington, D.C.

the opposition.[3] Ensuring such a capability implies protecting the network and the enabling information infrastructure, not only the information itself. Indeed, the profound importance of information assurance (IA) for network-centric operations is highlighted in the Department of Defense's (DOD's) *2006 Quadrennial Defense Review:* "Achieving the full potential of net-centricity requires viewing information as an enterprise asset to be shared and as a weapon system to be protected."[4] More fundamentally, there is increasing recognition that the very nature of network-centric operations—which implies the interconnection of everyone and everything—introduces threats and vulnerabilities, allowing many points of potentially harmful entry and paths for propagation of opposition attacks on information.

Indeed, the damaging effects of isolated domestic or international hackers on common commercial Internet grids are all too common; for example, a 2006 computer attack at the Naval War College forced the campus to shut down its connection to the Internet.[5] The impact then of a concerted attack by an enemy nation or state against U.S. computing and communications resources and infrastructure is not only potentially drastic in scope, but also increasingly more likely to occur. In this regard a key issue is the abundant use of vulnerable commercial off-the-shelf technologies, and further complicating this trend is the growing movement toward a homogeneous information system infrastructure, presenting one "target." Such realities present a threat to information assurance.

In recent years the Department of the Navy (DON) has established its "FORCEnet" vision as the Navy's approach to implementing network-centric operations.[6] This vision presents an operational view of capabilities, architectures, and concepts inclusive of the entire naval force—a view that is heavily dependent on the assured security and reliability of the Navy's information infrastructure. Also, the FORCEnet vision and systems for naval forces are both heavily integrated with and influenced by related information systems and networks across the entire DOD enterprise. The present study was motivated by this FORCEnet vision for naval network-centric operations, by recognition of the growing threats to information certainty, and by the need for better understanding and management of the many information assurance issues and influences both by naval forces and across the DOD. A basic premise of the study is the belief that in the FORCEnet/network-centric world of the DON and the DOD, information assurance cannot

---

[3]Naval Studies Board, National Research Council. 2000. *Network-Centric Naval Forces: A Transition Strategy for Enhancing Operational Capabilities,* National Academy Press, Washington, D.C.

[4]Department of Defense. 2006. *2006 Quadrennial Defense Review,* Washington, D.C., February.

[5]James Sherman. 2006. "Computer Attack Shuts Down Naval War College Networks," *Inside Defense,* Washington Defense Publishers, Washington, D.C., November 27.

[6]For additional background on FORCEnet, see National Research Council, 2005, *FORCEnet Implementation Strategy,* The National Academies Press, Washington D.C.; and National Research Council, 2006, *C4ISR for Future Naval Strike Groups,* The National Academies Press, Washington, D.C.

be treated as an isolated subject. Information assurance is not just about ensuring proper password practices, installing firewalls, and applying software patches, viewed in isolation from actual operations. Rather, information assurance as a critical requirement for operational success has to be fused with and subsumed into broader operational thinking, since the success of operations is the ultimate objective and measure of information assurance. Failure to accomplish information assurance would inevitably have a high negative impact on the ability of naval forces to achieve their missions.

TERMS OF REFERENCE

A letter dated December 7, 2007, from ADM Gary Roughead, Chief of Naval Operations, to Dr. Ralph J. Cicerone, President of the National Academy of Sciences, requested that the National Research Council's (NRC's) Naval Studies Board (NSB) conduct a comprehensive study on information assurance issues for U.S. naval forces. The purpose of the requested study was to review and address specific information assurance issues critical to network-centric naval operations, including vulnerabilities and potential mitigating actions that might be taken by the Department of the Navy.[7,8]

Accordingly, the National Research Council, under the auspices of its Naval Studies Board, established the Committee on Information Assurance for Network-Centric Naval Forces in February 2008.[9] The study's terms of reference, formulated by the Chief of Naval Operations' staff in consultation with the NSB chair and director, charge the committee to produce two reports over a 12-month period. First, after its second full meeting, the committee was to produce a letter report that did the following:

- Summarized the key information assurance initiatives underway within the Naval NETWAR/FORCEnet Enterprise,[10]
- Recommended any near-term information assurance needs for network-centric naval forces, and
- Identified defense-related efforts that the naval forces should take advantage of and/or ensure compatibility with.

[7]Acronyms and abbreviations are provided in Appendix A.

[8]The study's full terms of reference are provided in Appendix B.

[9]Biographical information for the committee members is presented in Appendix C.

[10]The Naval NETWAR/FORCEnet Enterprise includes the Office of Chief of Naval Operations; the Naval Network Warfare Command; the Space and Naval Warfare Systems Command; the Program Executive Office for Command, Control, Communications, Computers, Intelligence (C4I) and Space; and others who provide C4I and information operations support to the naval forces.

The committee was requested to produce a comprehensive final report, following the letter report, that addresses the full terms of reference. The requested letter report was delivered to the Chief of Naval Operations in November 2008 and was briefed to multiple constituents during which discussions were held on many of the immediate IA issues. This report—the committee's comprehensive final report—builds on the important areas identified in the letter report. The committee believes that it has responded productively and has provided a comprehensive analysis and solid recommendations for actions to help position network-centric naval forces for their continued mission assurance.

THE COMMITTEE'S APPROACH

In accomplishing its task, the committee took on a wide range of information assurance topics as requested in the terms of reference. The committee organized itself first to understand the nature of the naval information assurance issues and threats, then to understand current IA actions and responsibilites across both the DON and the DOD, and finally to formulate suggested IA responses and actions for naval forces that take into consideration operational, technical, and organizational viewpoints and needs. The findings and recommendations in this final report are based on wide-ranging input from experts and documents, both internal and external to naval operations and the DOD, and on the committee's own analysis, which draws on the expertise and experience of its members.

The committee was first convened in March 2008. After its first two meetings, the committee drafted its interim letter report. It held additional meetings and site visits over a period of 6 months, both to gather input from the relevant communities and to discuss its findings and recommendations. An outline of the committee's meetings is provided below:

- *March 5-6, 2008, in Washington, D.C. First full committee meeting. Briefings on information assurance issues, responsibilities, initiatives, strategies, and studies:* Office of the Deputy Chief of Naval Operations for Communications Networks; Office of the Deputy Department of the Navy Chief Information Officer; Information Assurance Directorate, Naval Network Warfare Command; Office of the Deputy Assistant Secretary of Defense for Information and Identity Assurance; Director, C4, and Chief Information Officer, U.S. Marine Corps; Office of the Department of the Navy Chief Information Officer; and Office of the Director, Information, Services and Integration; Secretary of the Air Force Office of Warfighting Integration and Chief Information Officer; Air Force Scientific Advisory Board; and Defense Science Board.
- *April 10, 2008, at Fort Meade, Maryland. Site visit. Briefings on information assurance initiatives and strategies:* National Security Agency, Information Assurance Directorate.

• *April 28-29, 2008, in Norfolk, Virginia. Second full committee meeting. Briefings on computer network defense, defense in depth, information assurance initiatives, Navy/Marine Corps Intranet, and naval information assurance strategies:* Naval Network Warfare Command (including Navy Cyber Defense Operations Command and Navy Global Network Operations and Security Center); and Network Systems Personnel—USS *Normandy* (CG-60).

• *May 29-30, 2008, in Washington, D.C.; Ashburn, Virginia; and Arlington, Virginia. Third full committee meeting. Briefings on the Next Generation Enterprise Network, the Consolidated Afloat Networks and Enterprise Services, and the Comprehensive National Cyber Initiative:* Office of the Deputy Chief of Naval Operations, Communications Networks; and Office of the Director of National Intelligence. *Site visit. Briefings on network security and information assurance commercial best practices:* Verizon Government Network Operations and Security Center. *Site visit. Briefings on DOD global network operations, cyberdefense, and information assurance initiatives:* Joint Task Force–Global Network Operations (JTF–GNO).

• *June 17-18, 2008, in Washington, D.C. Fourth full committee meeting. Briefings on information assurance/cyberdefense-related programs, studies, and research and development:* United States Marine Corps Network Operations and Security Command; Office of Information Assurance Division, Headquarters U.S. Marine Corps; Office of Deputy Chief of Naval Operations for Manpower, Personnel, Training and Education; the Defense Advanced Research Projects Agency; Office of Naval Research; the Naval Reseach Laboratory; and Office of Program Management, Program Executive Office (PEO) Ships.

• *July 16, 2008, at Fort Meade, Maryland. Follow-up site visit. Briefings on information assurance and cyberdefense-related initiatives:* National Security Agency, Information Assurance Directorate.

• *July 17-18, 2008, in Washington, D.C., and Arlington, Virginia. Fifth full committee meeting. Briefings on information assurance and cyberdefense-related initiatives, studies, and commercial best practices:* Chief of Naval Operations Strategic Studies Group; Computer Science and Telecommunications Board, the National Research Council; Office of the Chief Technology Officer, Defense Information Systems Agency; Office of Naval Intelligence; Citigroup Inc., IT Risk and Program Management; Verizon, Security Solutions Division; and Office of the Commander, U.S. Pacific Fleet.

• *August 4-5, 2008, in San Diego, California. Site visit. Discussion of IA-related issues, strategies, and initiatives:* U.S. Navy Space and Naval Warfare Systems Command, PEO-Command, Control, Communications, Computers, and Intelligence (C4I); and the Office of the Commander, U.S. Third Fleet.

• *August 18-22, 2008, in Woods Hole, Massachusetts. Sixth full committee meeting. Committee deliberations and report drafting.*

• *October 10, 2008, in Washington, D.C. Site visit.* Office of the Director, Naval Nuclear Propulsion Program.

The months between the committee's last meeting and the publication of the report were spent preparing the draft manuscript, gathering additonal information, reviewing and responding to the external review comments, editing the report, and conducting the security review needed to produce an unclassified report.

Acknowledgment of Reviewers

This report has been reviewed in draft form by individuals chosen for their diverse perspectives and technical expertise, in accordance with procedures approved by the National Research Council's Report Review Committee. The purpose of this independent review is to provide candid and critical comments that will assist the institution in making its published report as sound as possible and to ensure that the report meets institutional standards for objectivity, evidence, and responsiveness to the study charge. The review comments and draft manuscript remain confidential to protect the integrity of the deliberative process. We wish to thank the following individuals for their review of this report:

Brig "Chip" Elliott, BBN Technologies,
Carl E. Landwehr, McLean, Virginia,
Frank T. Leighton, Massachusetts Institute of Technology,
Dawn Meyerriecks, Purcellville, Virginia,
John E. Rhodes, LtGen, USMC (retired), Balboa, California,
Jonathan M. Smith, University of Pennsylvania,
William D. Smith, ADM, USN (retired), Fayetteville, Pennsylvania, and
William O. Studeman, ADM, USN (retired), Severna Park, Maryland.

Although the reviewers listed above provided many constructive comments and suggestions, they were not asked to endorse the conclusions or recommendations, nor did they see the final draft of the report before its release. The review of this report was overseen by Robert J. Hermann of Global Technology Partners, LLC. Appointed by the National Research Council, he was responsible for making certain that an independent examination of this report was carried out in accordance with institutional procedures and that all review comments were carefully considered. Responsibility for the final content of this report rests entirely with the authoring committee and the institution.

Contents

Summary

At the request of the Chief of Naval Operations, the Naval Studies Board, under the auspices of the National Research Council (NRC), established a committee to examine a wide set of issues associated with information assurance (IA) for network-centric naval forces.[1,2] Owing to the expansion of network-centric operating concepts across the Department of Defense (DOD) and the growing threat to information and cybersecurity from lone actors, groups of like-minded actors, nation-states, and malicious insiders, information assurance is an area of significant and growing importance and concern. Because of the forward positioning of both the Navy's afloat and the Marine Corps expeditionary forces, IA issues for naval forces are exacerbated, and are tightly linked to operational success. Broad-based IA success is viewed by the NRC's Committee on Information Assurance for Network-Centric Naval Forces as providing a central underpinning to the DOD's network-centric operational concept and the Department of the Navy's (DON's) FORCEnet operational vision.[3] Accordingly, this report provides

[1]The NRC's Committee on Information Assurance for Network-Centric Naval Forces first met in March 2008. The study's terms of reference are found in Appendix B. This report, the full final report from this study, follows the committee's interim letter report, dated November 6, 2008.

[2]During the course of its study, the committee held meetings in which it received (and discussed) materials that are exempt from release under 5 U.S.C. 552(b). A summary of the committee's meeting agendas is provided in the Preface of this report.

[3]FORCEnet is defined as "the operational construct and architectural framework for naval warfare in the information age that integrates warriors, sensors, networks, command and control, platforms, and weapons into a networked, distributed, combat force that is scalable across all levels of conflict from seabed to space and sea to land." See National Research Council, 2005, *FORCEnet Implementation Strategy*, The National Academies Press, Washington, D.C., p. 1.

a view and analysis of information assurance in the context of naval "mission assurance."

The growing threats to naval networks and computer systems coupled with the DON's increasing employment of commercial information technology (IT) as a critical part of warfighting systems require the DON to take significant action to reduce its current and emerging IA risks. This will require an IA strategy to guide the Navy and Marine Corps in defining and managing a broad array of interrelated IA activities. It will require that these activities be properly integrated to provide the basis for a naval IA risk management system that can respond to a continuously changing set of IA threats.

While the study identified many positive naval IA efforts currently underway, it also identified the following areas where new, coordinated IA-focused efforts will be required in order for the naval forces to achieve important levels of risk reduction:

- Doctrine development, operational procedures, and operational exercises to provide resilience against successful information system attacks;
- Technology research, development, and deployment—including system architecture research;
- Education and training for all naval personnel and the development of specialized career paths;
- Intelligence gathering and assessment;
- The IT acquisition process;
- Risk analysis methods for prioritizing investments;
- Dynamic and adaptive network and system reconfiguration; and
- Network and system monitoring.

The report addresses each of the above areas and related issues associated with information assurance and cyberdefense—issues that in many cases are very intertwined and have impact across the entire spectrum of DON and DOD enterprises. As such, the activities for reducing IA risk cut across many of the DON's current management domains and face serious organizational obstacles to achieving the needed integration of efforts.

Based on presentations to the study committee[4] and a review of available documentation related to naval and DOD IA, this report presents its case for action through a discussion of the following subjects: (1) the threats to IA, (2) the technology trends that contribute to potential IA and mission threats, and (3) a review of current DOD and DON IA initiatives deployed to help mitigate these trends and threats. The report then presents arguments for additional actions that the DON should undertake in its longer-term operational and technical response to IA-related mission threats—actions that the committee believes should begin

[4]See the Preface for a summary of the committee's data-gathering sessions.

immediately owing to the rapidly evolving nature of the threats and considering the time that will be required to mature and regularize the new approaches to IA that will result from the changes. Items such as Non-Classified Internet Protocol Router Network (NIPRnet) and Secret Internet Protocol Router Network (SIPRnet) security, elements of updated potential cyber concepts of operations (CONOPS; including the integration of offense-defense into cyber operations), the impact of network system architecture, advanced IA research and development (R&D) needs, IT acquisition, and cyber workforce development are all discussed in detail.

The report also presents evidence and discusses what are believed to be important shortfalls in the current naval approaches to IA-related risk management. On the basis of the identification of these shortfalls and the analysis of the surrounding IA issues, the committee presents a number of major findings and recommendations that offer necessary and practical approaches for improving matters.[5]

The DON's implementation of the committee's findings and recommendations would require a significant and sustained effort because of the breadth and nature of IA across the naval and DOD enterprise. The committee presents supporting evidence indicating that the likelihood of success on each of the report's recommendations would be greatly enhanced if the DON were to create an organizational structure that would allow the needed IA and related capabilities to be managed with clearer lines of responsibility and authority. The arguments and options for potential organizational changes are presented in the report's final chapter; these changes are recognized by the committee as being quite significant, but necessary to ensure long-term IA and network-centric operational success. The report suggests that a more centralized IA organizational construct be adopted, with clear responsibilities and authorities that cut across several existing IA governance seams.

PRIORITY AREAS FOR ACTION

The findings and recommendations in this final report build on the four findings and recommendations contained in the committee's interim letter report. Conclusions from this study can also be viewed in the context of the three general themes for recommended action presented below.

Action Area 1: Establish a Framework for Mission-Driven IA Risk Assessment

Presentations to the committee indicated that threats to IA are rapidly increasing. In addition, performance enhancements and economic opportunities made

[5]The chapters of this report contain additional important findings and recommendations as well as the 10 major findings and recommendations included in this Summary.

possible by more aggressive application of commercial IT are serving to increase the IA risks to mission execution being accepted by the Navy. It is not clear whether the trade-offs being made are purposeful or not, as there is little evidence of mission risk analyses accompanying the opportunity analyses for implementing new information system solutions. This study offers the following three major findings and recommendations that are related to this issue.

Update IA Operational Doctrine

Major Finding 1: Naval operations are highly dependent on information derived through all networks, including the Non-Classified Internet Protocol Router Network (NIPRnet) and legacy networks. The committee has seen evidence to suggest that the NIPRnet and legacy networks are highly vulnerable, and yet mission-critical functions such as managing logistics are being conducted on these shared networks.

Major Recommendation 1: To help address and reduce current perceived network risks related to the NIPRnet and legacy networks, the Department of the Navy should carry out the following:

- Undertake a systematic risk analysis to understand the mission impacts that could be created by information assurance failures. This analysis should be based on an understanding—derived through appropriate doctrinal, operational, procedural, and technical analyses—of the information and applications that reside on the networks and how they contribute to mission success.
- Evaluate the implementation of controls that balance operational security risks in posting information on the NIPRnet with the need for information sharing.
- Begin to design, architect, and implement the Department of the Navy's networks and systems with an objective of better separating the functions of mission-critical command-and-control systems, logistics, supply, and welfare and morale systems in such a way that an IA compromise in one of these functional areas does not create an IA compromise in others.
- Begin to develop IA operational doctrine that includes being able to conduct mission-critical operations with reduced information capabilities, minimize the time for restoration (reestablishing confidence in capabilities and data), and conduct training exercises for fighting through information attacks, including backup plans for the last mile of connectivity.[6]

[6]Major finding and recommendation 1 are found in the section entitled "Addressing NIPRnet and SIPRnet Threats" in Chapter 3 of this report.

Reexamine Network Separation Strategy for Critical Systems

Major Finding 2: The Global Information Grid (GIG) architecture promises to provide secure information services that are envisioned to be electronically integrated into weapons systems and other mission-critical control systems. This vision is highly dependent on trustworthy commercial off-the-shelf (COTS) technology components. The Department of the Navy, in keeping with the GIG architecture vision, is increasingly dependent on logical (software-based) information isolation rather than on physical separation for highly integrated, warfighting-critical systems composed largely of COTS components. This strategy is risky from an IA perspective, given the demonstrated vulnerabilities in COTS components.

Major Recommendation 2: The Office of the Assistant Secretary of the Navy for Research, Development and Acquisition (ASN[RDA]), in conjunction with other interested Navy and Marine Corps elements, should reexamine its IA architecture and design strategy, with emphasis on establishing the IA worthiness of the current systems under development. Special attention should be given to (1) the IA aspects of isolation and separation inherent in the outcomes in the Navy's Consolidated Afloat Networks and Enterprise Services (CANES) program and (2) the DDG-1000 onboard communication subsystem.[7,8]

Develop and Communicate IA Design Principles

Major Finding 3: As part of its implementation of network-centric warfare capabilities, the Department of the Navy is aggressively embracing integrative COTS technologies such as service-oriented architectures in order to take advantage of potential positive benefits, including wider information availability. However, these adaptations also have the potential to introduce new and possibly serious IA risks into naval systems. Unfortunately, existing naval systems do not appear to have been designed with consideration of the collateral IA risks as a foundational system attribute.

Major Recommendation 3: In order to provide the appropriate level of information assurance, the Office of the ASN(RDA) should adopt and manage system developments using sets of IA principles that are explicitly specified and required to be incorporated into the naval forces enterprise architecture, including specifically addressing the IA requirements of service-oriented architectures. In addition,

[7]Network design plans for the CANES program and the DDG-1000 (a planned new class of the U.S. Navy's multimission ships) are discussed in Chapters 2 and 3, respectively.

[8]Major finding and recommendation 2 are found in the subsection entitled "IA Risks of Current COTS Technologies" in Chapter 4.

these principles need to be embraced throughout the system life cycle and adopted by existing naval systems as they are upgraded.[9]

Action Area 2: Manage and Invest for Mission Assurance

Given the current trends related to the increasing vulnerability of information systems, naval forces face significant and growing risks of being unable to execute assigned missions. Reducing IA risks will require an integrated mixture of technological, procedural, and operational solutions to address possible enemy attacks. Potential solutions will include both enhanced defense to reduce the likelihood of successful attacks and enhanced resilience to attacks that are successful. Recognizing the range of possible attacks, efforts must be made to focus solution development to counter the set of attacks that are forecasted to be the most likely and would result in the most serious degradation of mission performance. This study offers the following four major findings and recommendations that are related to this issue.

Eliminate Shortcomings from Current IA Initiatives

Major Finding 4: The Department of the Navy has underway a diverse set of IA initiatives that are representative of best commercial IT practices. However:

- No means of integrated assessment exists for determining the impact of implementing the initiatives;
- The implementation of these initiatives will take significant resources and in some cases more than 3 years to implement, leaving a number of naval networks vulnerable to already-known exploitations; and
- Even if all of the existing initiatives are implemented and are successful, these networks are still not assured against the different and more sophisticated attacks that are likely to occur.

Major Recommendation 4: Because of the immediate and increasingly sophisticated nature of cyberthreats, the Office of the ASN(RDA), in collaboration with the Office of the Secretary of Defense and the National Security Agency, should conduct a thorough examination of technical opportunities and architectural options and develop a comprehensive plan for reengineering naval networks and computing enclaves to be resilient through cyberattacks by sophisticated adversaries. This plan needs to go beyond commercial best practices, incorporating advanced technology procedures that have been developed by DOD research agencies, mission assurance concepts, and active defense. The plan should also

[9]Major finding and recommendation 3 are found in the subsection entitled "Service-Oriented Architectures" in Chapter 4.

establish operational metrics, baseline these metrics, and set goals for their improvement.[10]

Improve Naval-Specific Cyberthreat Projections

Major Finding 5: The Navy has not comprehensively translated adversary capabilities into risk analysis assumptions or into an operational threat, and it does not routinely share the risk analyses and threat models that exist across the various Navy and Marine Corps organizations that have responsibility for information assurance. Based on the information briefed to the committee, there does not appear to be adequate emphasis on understanding how adversaries intend to or could use their capabilities and DOD network vulnerabilities to disrupt naval operations.

Major Recommendation 5: The Director, Naval Intelligence, in collaboration with the Defense Intelligence Agency and national intelligence organizations, should support cyber risk analysis by collecting and analyzing all source intelligence to improve the Department of the Navy's understanding of adversaries' mission intent, strategy, and tactics and to illuminate how these could impact the ability of the Navy and Marine Corps to accomplish their missions and objectives.[11]

Also, threat and risk analysis, specifically including CONOPS and operational capabilities of adversaries, should be shared across the many Navy and Marine Corps organizations with significant dependencies on information assurance. Standard scenarios and measures of effectiveness should be used by organizations responsible for information assurance.

Improve the IT Acquisition Process

Major Finding 6: Cyberthreats change on a timescale much shorter than the DOD acquisition life cycle for developing and deploying cybersecurity technologies. There are increasing risks from these cyberthreats, including risks of being unable to respond to assigned warfighting missions. Rapid acquisition and fielding of IA solutions are critical, but the committee did not see processes being put into place to support this need.

Major Recommendation 6: The committee recommends that the following specific actions be undertaken by the ASN(RDA), with the support of the Direc-

[10]Major finding and recommendation 4 are found in the section entitled "Summary Assessment of Initiatives" in Chapter 2.

[11]Major finding and recommendation 5 are found in the section entitled "Findings and Recommendations" in Chapter 5.

tor, Naval Research, to address the timely acquisition and implementation of IA solutions:

- Actively participate in DOD efforts to define and establish intelligence that provides predictions about future cyberattack techniques which are sufficient to stimulate development of defensive responses,
- Use existing operations and maintenance processes supplemented by design and prototyping activities carried out by naval laboratories to more rapidly develop and implement solutions,
- Establish a rapid technology testing and evaluation laboratory and a technology insertion program—modeled after the Future Naval Capabilities program—to leverage and accelerate ongoing research in cybersecurity into Navy networks, and
- Establish a standard management process styled after the urgent-need process for the Global War on Terrorism (as defined in SECNAV [Secretary of the Navy] Note 5000 on "Rapid Development and Deployment Response to Urgent Global War on Terrorism Needs").[12]

Increase Naval IA R&D Funding

Major Finding 7: The Department of the Navy has not established a sufficiently robust research program in IA. The funding level requested by the Office of Naval Research (ONR), approximately $2 million per year, is inadequate even to ensure that the DON effectively leverages the research investments of other agencies. Current gaps in information assurance capability for naval forces are made even more significant by a lack of strategy for investing in advanced R&D to redress these gaps.

Major Recommendation 7: The Director, Naval Research, should develop—and the Chief of Naval Operations (CNO) and the Commandant of the Marine Corps (CMC) should ensure funding for—a robust science and technology research program in information assurance. An order-of-magnitude increase in funding levels through ONR's Naval Research Laboratory would establish the Navy as a full participant in IA technology R&D, providing the knowledge base to guide and prioritize naval implementation choices and allowing the Navy to draw from the work of outstanding members of the academic and industrial research communities. The Navy should focus its research efforts on addressing capability gaps specifically related to the needs of naval forces that are not being sufficiently addressed elsewhere.

Concurrently, the Office of Naval Research should develop a rapid technology

[12]Major finding and recommendation 6 are found in the subsection entitled "Existing Naval Research and Development and Acquisition Processes" in Chapter 4.

insertion program to enable the rapid deployment of solutions for responding to new threats, based on both the leveraging of internal Navy research results and the use of ongoing research results derived from the funding of other R&D organizations, such as at the Defense Advanced Research Projects Agency, National Security Agency, Army Research Office, Air Force Office of Scientific Research, National Science Foundation, Department of Energy, and Department of Homeland Security.[13]

Action Area 3: Rethink IA—Suggested Doctrinal and Organizational Responses

The range of activities required to reduce the growing IA risks is very broad, involving the application of new technology and new operational doctrine. This range of activities is based on risk assessments that cut across the various naval missions and organizations, and they must be accomplished in coordination with the broader DOD activities addressing IA. In particular, IA cannot be treated in isolation, but rather must be considered in the broader context of military operations.

Recognizing that IA requires addressing the "weakest links" in the overall naval system of systems, a prioritization of IA enhancement activities is critical. Recognizing the speed with which new attacks can be designed, developed, and propagated, rapid-response solutions inserted into practice are required. The committee believes that new approaches are required for addressing naval IA needs into the future. It offers the following three major findings and recommendations related to this issue.

Develop Doctrine for Offense-Defense Integration

Major Finding 8: The four cyberspace IA-related domains of protecting, exploiting, attacking, and intelligence do not appear to be closely integrated in the Navy. In particular, the Department of the Navy does not appear to be aggressively considering and assessing alternatives to gain greater IA advantages through such integration.

Major Recommendation 8: The Office of the CNO and the Office of the CMC should consider approaches for reducing the separation and enhancing the integration across emerging offense, defense, and intelligence organizations related to IA.[14]

---

[13]Major finding and recommendation 7 are found in the subsection entitled "Current Naval Information Assurance Research and Development Budget" in Chapter 4.

[14]Major finding and recommendation 8 are found in the section entitled "Integrating Cyber Operations" in Chapter 3.

Update the Department of the Navy Cyber Workforce Strategy

Major Finding 9: The Department of the Navy's workforce, consisting of officers, enlisted personnel, and civilians, has not been required to possess a uniform, prerequisite set of knowledge and IT-related experience. Today's IA-related threats and trends point to a need for the Navy and Marine Corps to address education, training, and career paths as part of the needed response to the growing IA risks and the growing importance of naval cyber operations. The Navy's Corry Station cyber operations training program provides a strong and positive start toward meeting this need.[15]

Major Recommendation 9: The Office of the CNO and the Office of the CMC should establish a dedicated cyber workforce strategy to include all elements of personnel management (accession, reenlistment, retention, and assignment). Since cyber-related technology continues to evolve rapidly, the cyber workforce program for naval forces should also include measures to continuously modernize the Navy and Marine Corps training and education curriculum, including the development of formal relationships with universities and external advisers for guiding and supporting naval needs in cyber education and training.[16]

Adopt New Naval IA Organizational Structure

Major Finding 10: The governance of information assurance is widely distributed across naval forces, with many parties playing roles, resulting in many governance seams. In particular, there is no centralized authority or organizational mechanism in place in the Department of the Navy for governing IA and end-to-end cyber operations. For example, a shared scope of governance of security policy and fiscal authority for naval networks resides throughout the DON, including with the Department of the Navy Chief Information Officer; the Deputy CNO for Network Operations; Headquarters, Marine Corps; Naval Network Warfare Command; Echelon II Chief Information Officers; Commander–Naval Installation Command; Program Executive Officers; and Navy Systems Command.

Major Recommendation 10: The leadership of the Department of the Navy should examine more-centralized IA-related organizational structures for integrating its information assurance strategies and plans across all naval communities (surface, subsurface, expeditionary, air, space, and cyberspace), as well as for integrating those same strategies and plans with joint communities (Combatant

[15]The Navy's Corry Station cyber operations training program, operated as part of the Center for Information Dominance at Corry Station, is discussed in the subsection entitled "Career Paths" in Chapter 3.

[16]Major finding and recommendation 9 are found in the subsection entitled "Career Paths" in Chapter 3.

Command, Office of the Secretary of Defense). The examination should address the needed IA governance and fiscal authorities for sustaining both current and future readiness levels, as well as which DON organizations are critical to defending against evolving cyberthreats—from the strategic to the tactical level.[17]

While cost considerations were explicitly excluded from the committee's terms of reference, cost implications are an obvious consideration for addressing many of the findings and recommendations presented above. However, the committee believes that several of the major recommendations can be acted on with minimal additional capital or operating expenditures. Owing to the immediacy of the issues involved with information assurance for naval forces, the committee urges the consideration of all recommendations in a timely fashion.

[17]Major finding and recommendation 10 are found in the "Summary Discussion" of the subsection entitled "Alternative Organizational Models" in Chapter 6.

1

Background—Naval Network-Centric Operations, Information Assurance, and Current Cyberthreats

NETWORK-CENTRIC OPERATION AND ITS DEPENDENCIES

Multiple definitions exist for the term "network-centric," all being largely equivalent. To be specific, in this study the National Research Council's (NRC's) Committee on Information Assurance for Network-Centric Naval Forces adopts the following definition from prior NRC reports conducted under the auspices of the Naval Studies Board (NSB):

> Network-centric operations are military operations that exploit state-of-the-art information and networking technology to integrate widely dispersed human decision makers, situational and targeting sensors, and forces and weapons into a highly adaptive, comprehensive system to achieve unprecedented mission effectiveness.[1,2]

The NSB's report *Network-Centric Naval Forces* further characterizes network-centric operations in the following manner:

> Forward deployment of naval forces that may be widely dispersed geographically, the use of fire and forces massed rapidly from great distances at decisive

[1]Naval Studies Board, National Research Council, 2000, *Network-Centric Naval Forces: A Transition Strategy for Enhancing Operational Capabilities,* National Academy Press, Washington, D.C., p. 1.

[2]For additional reading on this topic, see National Research Council, 2006, *C4ISR for Future Naval Strike Groups,* The National Academies Press, Washington, D.C., pp. 36-37; and National Research Council, 2005, *FORCEnet Implementation Strategy,* The National Academies Press, Washington, D.C., p. ix.

locations and times, and the dispersed, highly mobile operations of Marine Corps units are examples of future tasks that will place significant demands on networked forces and information superiority. Future naval forces must be supported by a shared, consolidated picture of the situation, distributed collaborative planning, and battle-space control capabilities. In addition, the forces must be capable of coordinating and massing for land attacks and of employing multi-sensor networking and targeting for undersea warfare and missile defense.[3]

The idea of network-centric operations[4] has become centrally embedded in naval concepts and plans for operations. This is manifested, for example, in the stand-up of the Naval Network Warfare Command and the evolution of the Marine Corps Network Operations and Security Command. It is also apparent in the development and use of the FORCEnet concept; the program priorities of the Office of the Chief of Naval Operations (N6) and Marine Corps; program development by the Assistant Secretary of the Navy for Research, Development and Acquisition (ASN[RDA]); and experiments conducted in the Trident Warrior experimentation program.

Since network-centric operations involve, for example, the synchronized execution of distributed operations and the widespread sharing of situational awareness and decision-making data, they require a dependable underlying information and communications infrastructure. This requirement is made explicit in the three goals for network-centric operations that the Assistant Secretary of Defense for Networks and Information Integration has established for the entire Department of Defense (DOD):

Goal #1—Make information available on a network that people depend on and trust.

Goal #2—Populate the network with new, dynamic sources of information to defeat the enemy.

Goal #3—Deny the enemy comparable advantages and exploit weaknesses.[5]

FORCEnet can be regarded as the naval means for achieving the goals listed above. It is envisioned by the Navy and Marine Corps as the naval element of

---

[3]Naval Studies Board, National Research Council, 2000, *Network-Centric Naval Forces: A Transition Strategy for Enhancing Operational Capabilities,* National Academy Press, Washington, D.C., p. 12.

[4]The Department of Defense uses the term "net-centric" rather than "network-centric" in its more current documents. For the sake of editorial consistency, this report will use the term "network-centric" as it first appeared publicly in a 1998 *U.S. Naval Institute Proceedings* article entitled "Network-Centric Warfare: Its Origin and Future," January, by VADM Arthur K. Cebrowski, USN, and John Gartska.

[5]Written statement by Lt Gen Charles E. Crooms, Jr., USAF, Director, Defense Information Systems Agency, before the U.S. House Armed Services Committee, April 6, 2006. Available at <http://www.globalsecurity.org/military/library/congress/2006_hr/060406-croom.pdf>. Accessed November 11, 2008.

the Global Information Grid (GIG) jointly existing on the GIG with other non-FORCEnet elements. This concept envisions that naval forces will be an integral part of a much larger joint, coalition-based, interagency and commercial network that will enjoy magnified support opportunities from the network because of its expanded scope. Within the GIG, naval nodes will be tightly integrated with non-naval nodes. Naval nodes will rely on information and services provided by non-naval elements, just as they will contribute uniquely naval capabilities to the wider GIG.

The following examples of network-centric operations make explicit their dependency on a dependable underlying information and communications infrastructure:

- *Synchronized execution of operations—depends on connectivity among distributed force elements.* Examples of such operations are those executed according to the Marine Corps concept for distributed operation of small units.[6]
- *Situational awareness drawing from distributed sensors—depends on connectivity for data access and on the integrity of those data.* An example of such situational awareness is the air and undersea "picture" maintained by naval strike groups.
- *Combat system operation responsive to the command-and-control system—depends on fault-free operation of hardware and software and on the integrity of data.* An example of such a combat system would be that controlling the defensive missiles aboard an Aegis cruiser.
- *Distributed, collaborative planning—depends on connectivity for collaboration among command elements and for access to data and services to develop courses of action, and on the integrity of those data and services.* An example of such planning would be that conducted for naval forces as part of joint operations in regional warfare (e.g., scenarios that might occur in Iraq or Afghanistan).
- *Supporting data drawn from a large variety of distant sources—depend on reach-back connectivity to the continental United States and other distant locations, and on the integrity of data received.* An example of such data would be intelligence, surveillance, and reconnaissance data collected by national means.

The disruption or denial of computation or communications connectivity and the corruption or destruction of data would highly degrade or even render ineffective the network-centric approach to operations. The greater the dependence on information-sharing and communications capabilities, the more attractive become attacks against them—by both highly sophisticated and less sophisticated adversaries—to undermine U.S. operations.

As a result, information assurance (IA), provided by protecting information and communications systems against the threats of adversaries, is seen as a vital

[6]Commandant, U.S. Marine Corps (Gen Michael W. Hagee, USMC). 2005. *A Concept for Distributed Operations,* Headquarters, U.S. Marine Corps, Washington, D.C., April 25.

part of network-centric warfighting capabilities.[7] The FORCEnet Functional Concept states this need as follows:

> FORCEnet must therefore include the capability to protect command and control activities against efforts to deceive, exploit or otherwise attack them. This capability should include the abilities to detect, locate, and identify hostile information operations, defeat or counter those efforts, and mitigate the effects of successful hostile efforts. Information assurance also applies to accidental corruption of information. It should include the ability to recover to an earlier information state from any kind of information corruption.[8]

Both the current and future potential threats that must be confronted to realize these objectives and thereby ensure the successful execution of network-centric modes of operation are substantial, as the next section describes. Box 1.1 describes the unique naval mission element requirements of Sea Strike, Sea Shield, Sea Basing, Expeditionary Maneuver Warfare, and Sea Warrior, Sea Enterprise, and Sea Trial as they relate to naval forces' IA.

NATURE OF THE CYBERTHREAT

The cybersecurity threat environment, in terms of possible attack techniques, is effectively limitless. Many malicious exploits have been identified that have taken advantage of military information systems environments. Comprehensive implementation of information assurance practices must protect against a significant portfolio of potential threats. This section describes in a manner appropriate for public release the understanding of the threat developed by the committee.

Broad Categorization of Threat Types

At the top-most level, the cyberthreat can be broken into four types: as described below, they involve remote access, close access, life-cycle or supply chain insertion, and insiders. The intended purpose of these threats is to disrupt system functions (e.g., degrading or denying communications connectivity), to modify data (e.g., corrupting or falsifying data), and/or to steal data.

1. *Remote access.* Remote access refers to penetrations of or other disruptive actions to an information system gained through that system's connectivity

---

[7] In the committee's work, cybersecurity vulnerability and information assurance vulnerability are viewed as inseparable and are therefore treated in this report as equivalent.

[8] ADM Vern Clark, USN, Chief of Naval Operations; and Gen Michael W. Hagee, USMC, Commandant of the Marine Corps. 2002. *FORCEnet: A Functional Concept for the 21st Century,* Department of the Navy, Washington, D.C., February 2. Available at <http://www.navy.mil/navydata/policy/forcenet/forcenet21.pdf>. Accessed November 10, 2008.

BOX 1.1
Naval Missions and Information Assurance: A FORCEnet Viewpoint

Operationally, FORCEnet refers to the systems and processes for providing effective networked naval command and control in the 2015-2020 time frame. Command and control constitute the means and methods by which a commander recognizes what needs to be done in any given situation and sees that appropriate actions are taken. Every area of naval warfare, as described in the Naval Operating Concept for Joint Operations, Naval Power 21, Sea Power 21, and Marine Corps Strategy 21 will require FORCEnet to provide command-and-control functionality as follows:

- *Sea Strike:* FORCEnet will provide synchronization of distributed strike and assessment assets for Sea Strike's projection of offensive power from the sea. The collection, integration, and dissemination of surveillance, targeting, planning, and assessment information will facilitate the decision-making process through real-time collaborative planning and intelligent decision aids. FORCEnet will support the Joint Task Force Commander's task of coordinating and controlling the tempo and effects of complex and simultaneous joint assets and events. FORCEnet will enable the Commander to select and apply the most appropriate tactic and system to achieve the desired effect, whether kinetic, nonkinetic, strategic, operational, or tactical.
- *Sea Shield:* FORCEnet will enhance naval contributions to homeland defense and support assured access for joint, allied, and coalition forces overseas. Through capabilities provided by FORCEnet, Sea Shield will defend the sea battlespace and project defensive power from the sea over friendly forces ashore. FORCEnet will provide a common, integrated, user-tailored, and real-time operational picture coupled with rapid combat identification and near-real-time speed of command. Real-time collaboration and intelligent decision aids will complement all aspects of Sea Shield. FORCEnet will allow for threat engagements beyond a single platform's organic capability, and will allow carrier and expeditionary strike groups to act as single integrated and distributed combat systems.
- *Sea Basing:* Sea Basing increases the operational maneuver space and independence of naval and joint forces, improves speed of maneuver and reconstitution, and facilitates personnel and logistics sustainment functions without vulnerable shore footprints. FORCEnet's robust collaboration and planning capabilities and the seamless flow of large volumes of secure information supporting readiness, total asset

visibility, and sustainment will be key benefits to Sea Basing. FORCEnet capabilities will significantly enhance the ability of Marine forces to conduct Expeditionary Maneuver Warfare, Operational Maneuver from the Sea, and Ship to Objective Maneuver from a sea base. FORCEnet will allow joint commanders to exercise command and control in secure and mobile facilities, while allowing forces to arrive and be sustained on scene at maximum possible readiness. FORCEnet will yield access to information and total visibility and speed of delivery to Sea Basing activities for all classes of readiness and sustainment support.

- *Expeditionary Maneuver Warfare:* FORCEnet allows for collaborative planning while en route to and closing on objectives. FORCEnet will allow deployed forces to exchange critical information with other U.S., allied, and coalition forces during joint and combined operations. During ship-to-shore movements, forces that are virtually connected to the platforms from which they were launched, other forward-deployed forces, and distant sites will collect and share intelligence data for current and future operations. Forces will gain tremendous advantage through more rapid collection and dissemination of information, enabling more rapid and decisive decision making during sustained operations ashore. FORCEnet will incorporate appropriate capabilities from the Expeditionary Maneuver Warfare Capabilities List. FORCEnet will allow Marine forces to serve as the nucleus of, and provide an operating force for, a Joint Task Force Headquarters.
- *Sea Warrior, Sea Enterprise, and Sea Trial:* FORCEnet's robust, collaborative, information sharing, distributed services, and decision superiority benefits will also extend to the non-warfighting enterprise domain. FORCEnet provides Sea Warrior with near-real-time information services for personnel and personnel management, training, medical support, professional growth, and other personnel considerations. FORCEnet provides Sea Enterprise with the ability to transform business and financial processes and to produce essential infrastructure efficiencies. FORCEnet extends to Sea Trial a shared and time-sensitive environment in which to collaborate and validate new concepts and technologies.

SOURCE: Department of the Navy Enterprise Architecture Management View. Available at <http://www.doncio.navy.mil/EATool/Forcenet/Forcenet_home.htm#description>. Accessed February 27, 2009.

to a publicly accessible network (e.g., the Internet). An example of these remote access operations would be ones conducted against systems on the Non-Classified Internet Protocol Router Network (NIPRnet). Depending on the techniques used, these operations could gain access to a limited set of the system resources (e.g., files owned by one user) or to all the resources on a local area network (e.g., those controlled by a system administrator). Short of actual penetration, the operations could cause a degradation of network connectivity (i.e., denial of service) either by flooding the interfaces to the external networks with large amounts of network traffic or by disabling the operation of some intermediate network components (e.g., routers). Perpetrators of these remote access operations run all the way from "script kiddies"[9] through criminals and terrorists to world-class nation-state adversaries.

The number of attempted remote penetrations of U.S. government and naval systems has escalated over the past few years. The committee has had access to data and briefings indicating that these attempted intrusions into government and private networks have also become more sophisticated and more malicious. Such tactics as targeted "spear-phishing"[10] are now a common occurrence.

Remote access operations are the most commonly discussed means of penetration or other degradation, probably because they are the most visible. That does not mean, however, that the other means of penetration may not have consequences that are just as serious, if not more so.

2. *Close access.* Close access refers to penetrations effected against "closed" (typically classified) systems—that is, those not directly accessible through public networks. The Secret Internet Protocol Router Network (SIPRnet) would be an example of such a network. Close access could be achieved through direct physical tapping established through human or mechanical means, or through electromagnetic interaction with the closed system. Access to these "closed" systems might also be possible through remote means that exploit software vulnerabilities, as such systems may only be logically, not physically, separated from the public networks. Historically, the DOD has paid more attention to the detection of remote access penetrations than it has to close access detections, since the "closed" systems were felt to be safe by virtue of their physical and cryptographic isolation. Recently, for reasons discussed later, the DOD has begun to pay more attention to the possibility of close access penetrations.

---

[9]"Script kiddie" is a term applied to an amateur hacker, typically one seeking opportunist exploits.

[10]"Spear-phishing" is an e-mail spoofing fraud attempt that targets a specific organization, seeking unauthorized access to confidential data. As with the e-mail messages used in regular phishing expeditions, spear-phishing messages appear to come from a trusted source. Phishing messages usually appear to come from a large and well-known company or Web site with a broad membership base, such as eBay or PayPal. In the case of spear-phishing, however, the apparent source of the e-mail is likely to be an individual within the recipient's own organization and generally someone in a position of authority.

3. *Life-cycle (or supply chain) insertion.* Life-cycle insertion is the surreptitious insertion of modified hardware or software components into network components and information systems during their manufacture or maintenance.[11] The purpose of the inserted components would be to provide "back doors" for clandestinely exfiltrating information or, on receiving some sort of cue, disrupting the operation of the networks or information systems. These risks stem from the fact that potential adversaries play a key role in the offshore development and life-cycle support of commercial off-the-shelf (COTS) technology components[12] that are a critical part of the DOD's information architecture. Such activities have provided the basis for actual cases of embedding disabling technologies as part of seemingly normal technology products.

This risk is exacerbated by certain adversaries who have the necessary design skills to embed disabling technologies in ways that are extremely difficult to discover and who are able to incorporate disabling technology updates at the normal and rapid rates of product enhancement. Life-cycle insertion activities thus pose a serious threat because they are beyond the hypothetical and, if applied in certain operational circumstances, can significantly reduce U.S. military warfighting capability.

4. *Insiders.* Insiders are individuals within an organization who have access to its information systems and networks and who act in some way to the detriment of the system. They range from legitimate users who carry out harmful acts inadvertently to individuals who act with highly malicious intent. An inadvertent user could be one who, unknown to that individual, inserts a memory stick containing "malware" that would allow a compromise of the information system and associated network, potentially including "closed" networks. Instances of such activities have been regularly reported.[13]

A malicious user could be one recruited by a foreign intelligence agency or other adversarial party who would provide that agency or party with access to the network. In the worst case, this recruited insider would be one who has special knowledge of the technical details of the network or the information held on it and who passes that information on to a foreign intelligence agency. Recently,

---

[11]Samuel T. King, Joseph Tucek, Anthony Cozzie, Chris Grier, Weihang Jiang, and Yuanyuan Zhou. 2008. "Designing and Implementing Malicious Hardware," *Proceedings of the First USENIX Workshop on Large-Scale Exploits and Emergent Threats (LEET)*, San Francisco, Calif., April. Also available at <http://www.usenix.org/events/leet08/tech/full_papers/king/king_html/>. Accessed February 18, 2008. See also, Defense Science Board, 2007, *Mission Impact of Foreign Influence on DOD Software,* Office of the Under Secretary of Defense for Acquisition, Technology, and Logistics, Washington, D.C., September.

[12]The committee defines commercial off-the-shelf (COTS) technology to include commercial open-source developments.

[13]For example, see Bill Whitney and Tara Flynn Condon, 2008, "Five Ways Insiders Exploit Your Network," *NetworkWorld,* May, at <http://www.computerworld.com/action/article.do?command=viewArticleBasic&articleId=9083978>. Accessed November 10, 2008.

the DOD has emphasized the need for increased counterintelligence activities to protect against this class of threat.

Examples of Cyberthreats

Commercially available cybersecurity tools are predominantly reactive in the sense that they are used to address known vulnerabilities and threats that have been identified and characterized. Security patches are a major part of the current reactive response process. Patches are developed and deployed to address vulnerabilities that have been exploited and identified, but do not address zero-day attacks.[14]

Exploits that are "noisy" are relatively easy to identify. Increasingly, exploits are being discovered that are "quiet" by design, as the motivation for malicious code has moved to hacking for money and to running covert operations for gaining intelligence. As a result, well-resourced teams of engineers are designing, implementing, and vigorously testing malicious codes prior to releasing them, not unlike well-funded commercial software development firms.[15] These threats are very difficult to discover because they are engineered to live in harmony with the host while evading host-level sensors.

Figure 1.1 provides some examples of cyberthreats. As seen in the figure, these threats and their variants are growing rapidly. No limiting factor has been identified that can be expected to "cap" the threat environment. As discussed above, commercial technology responses to these threats are primarily reactive and hence, at best, can barely keep up with the advancing threats. The situation for the Department of the Navy (DON) is worse because its technology deployment processes are generally slower than those of commercial industry.[16]

The preceding observations are summarized in the following finding.

FINDING: Cyberthreats change on a timescale much shorter than the typical Department of Defense acquisition life cycle for developing and deploying

[14]A "zero-day" attack takes advantage of targeted computer application vulnerabilities before a patch has been created or applied. It is named zero-day because it occurs before the first day the vulnerability is disclosed.

[15]Samuel T. King, Joseph Tucek, Anthony Cozzie, Chris Grier, Weihang Jiang, and Yuanyuan Zhou. 2008. "Designing and Implementing Malicious Hardware," *Proceedings of the First USENIX Workshop on Large-Scale Exploits and Emergent Threats (LEET)*, San Francisco, Calif., April. Also available at <http://www.usenix.org/events/leet08/tech/full_papers/king/king_html/>. Accessed February 18, 2008.

[16]For example, a 2007 report from the Navy's Program Executive Office for Command, Control, Communications, Computers and Intelligence (PEO C4I) states that the average age for Navy C4I networks is 6.7 years, and the average time to market for new capabilities is 2 to 3 years. See <http://www.afcea-sd.org/C4ISR2007SymposiumArchive/C4ISRDownloads/2007C4ISRPresentations/Day%202/Day%20PM%20Keynote/070523_AFCEA_Symposium_FINAL.ppt>. Accessed February 26, 2009.

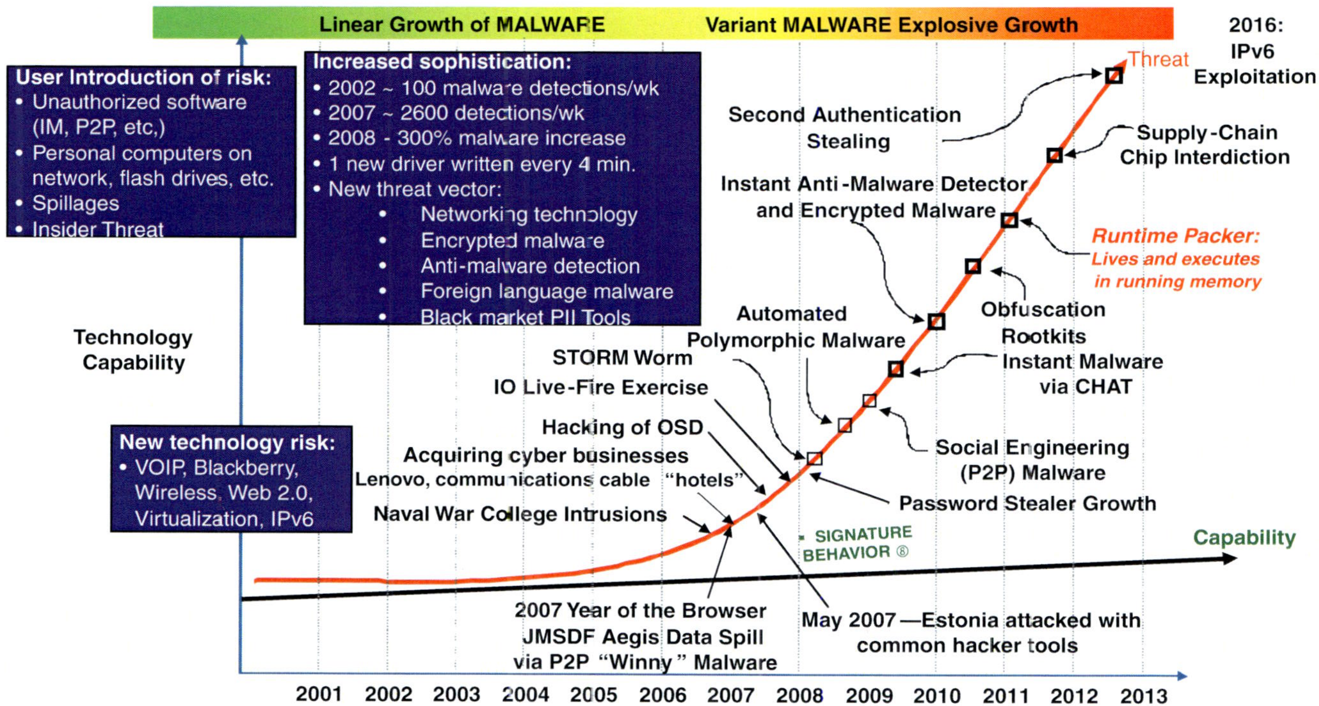

FIGURE 1.1 Trends in the growing quantity and sophistication of adversaries' cyberthreats and cyberattacks. NOTE: Acronyms are defined in Appendix A. SOURCE: RADM(S) David G. Simpson, USN, Director, Navy Networks, Deputy Chief of Naval Operations, Communication Networks (N6), "Next Generation Enterprise Network (NGEN) and Consolidated Afloat Networks and Enterprise Services (CANES)," presentation to the committee, May 29, 2008, Washington, D.C.

cybersecurity technologies. Several trends presented to the committee point to continuously increasing risks from these threats. Because the Navy is increasingly conducting warfighting using commercial information technology systems, these cyberthreats represent a serious threat to the Navy's warfighting capability.

Employment of Cyberattacks by Potential Adversaries

Reports of computer network intrusions by various adversaries continue to increase. Likewise, estimates of the number of adversary nation-states and other bodies (e.g., terrorists) skilled in the necessary computer technology to conduct intrusions are also increasing.[17] Significant among the reports of intrusions are numerous penetrations of networks owned by the U.S. government. Although these intrusions may not explicitly be attacks (i.e., they may not lead to damage or destruction of information or network resources), they require the same expertise and techniques required for computer network attack, including denial-of-service and data-corruption attacks.

Attribution of computer network intrusions is difficult, and it is very hard to be sure if any particular intrusion was conducted by a particular foreign government or other adversarial party. Nonetheless, special attention is currently being paid to the People's Republic of China (PRC). The *Annual Report to Congress: Military Power of the People's Republic of China, 2008,* prepared by the Office of the Secretary of Defense, states the following:

> In the past year, numerous computer networks around the world, including those owned by the U.S. Government, were subject to intrusions that appear to have originated within the PRC. These intrusions require many of the skills and capabilities that would also be required for computer network attack. Although it is unclear if these intrusions were conducted by, or with the endorsement of, the PLA [People's Liberation Army] or other elements of the PRC government, developing capabilities for cyberwarfare is consistent with authoritative PLA writings on this subject.
>
> • In 2007, the Department of Defense, other U.S. Government agencies and departments, and defense-related think tanks and contractors experienced multiple computer network intrusions, many of which appeared to originate in the PRC.
>
> • Hans Elmar Remberg, Vice President of the German Office for the Protection of the Constitution (Germany's domestic intelligence agency), publicly accused China of sponsoring computer network intrusions "almost daily." Remberg stated, "Across the world the PRC is intensively gathering political, military, corporate-strategic and scientific information in order to bridge their [sic] technological gaps as quickly as possible." Referring to reports of PRC

[17]John Rollins and Clay Wilson. 2007. *Terrorist Capabilities for Cyber Attack: Overview and Policy Issues,* Congressional Research Service, Washington, D.C., January 22. Available at <http://www.fas.org/sgp/crs/terror/RL33123.pdf>. Accessed February 11, 2009.

infiltration of computer networks of the German government, German Chancellor Angela Merkel said, "we must together respect a set of game rules." Similarly, in September 2007, French Secretary-General of National Defense Francis Delon confirmed that government information systems had been the target of attacks from the PRC.

- In addition to governments, apparent PRC origin network intrusions targeted businesses. In November 2007, Jonathan Evans, Director-General of the British intelligence service, MI 5, alerted 300 financial institution officials that they were the target of state-sponsored computer network exploitation from the PRC.[18]

Cybersecurity vulnerabilities are necessitating the introduction of entirely new warfighting doctrine. This fact is illustrated by the following extract on Chinese thinking from *Air Force and the Cyberspace Mission: Defending the Air Force's Computer Network in the Future*:

> China's ability to wage cyberwar against the United States is no longer speculation; it occurs daily and is growing exponentially. Two Chinese colonels wrote a paper in 2002 titled Unrestricted Warfare, wherein they candidly proposed using cyber attack as a new form of warfare against the United States. In their paper, they analyze United States military power and assess operations over the past decades and conclude "today, the independent use of individual technologies is now becoming more and more imaginable. The emergence of information technology has presented endless possibilities for match-ups involving old and new technologies and among new and advanced technologies."[19]

An important set of recent events involving Russia, Estonia, and Georgia also provide visibility with respect to the possibilities of more aggressive uses of cyberattacks as a complement to other elements of nation-state conflicts. Three cyberattack methodologies used during these events were reported in the press: first, the use of denial-of-service attacks to complicate the ability for adversaries to respond to a situation; second, through the use of the Internet, the rapid voluntary recruitment of participants to contribute to cyberattacks; and third, taking advantage of the confusion surrounding these activities, which makes it both complicated and time-consuming to accurately assess what is really happening, including attribution.

While the degree of accuracy of the above events in the specific press reports

---

[18]Office of the Secretary of Defense. 2008. *Annual Report to Congress: Military Power of the People's Republic of China, 2008,* Washington, D.C., pp. 3-4.

[19]Shane P. Courville, Lt Col, USAF. 2007. *Air Force and the Cyberspace Mission: Defending the Air Force's Computer Network in the Future,* Occasional Paper No. 63, Air War College, Center for Strategy and Technology, Maxwell Air Force Base, Ala., December. Available at <http://www.au.af.mil/au/awc/awcgate/awccsat.htm>. Accessed November 10, 2008.

can be argued,[20,21] each of the three uses is available to potential combatants, and the degree of use can certainly be escalated without incurring major costs or requiring long buildup times. Consequently, the committee recognizes that while the full-throttle use of these techniques has not yet been experienced, preparation for significant situations involving such methodologies is nonetheless necessary.

Need for Enhanced Analysis of Future Threats

It is well understood that the development of naval platforms must be supported by projections of future physical threats to those platforms (e.g., antiship missiles, undersea detection). Such threat projections are routinely provided by naval intelligence and the larger intelligence community. Similarly, projections of future cyberthreats are required for the development of platforms and information systems. All presentations to the committee on the subject of cyberthreat, however, focused almost exclusively on the current threat, apart from a few general examples of projected future threats (see, e.g., Figure 1.1).

The committee discussed this absence of future threat projections with representatives from program and acquisition management offices who briefed it. These representatives indicated that cyberthreat projections were absent at a level of detail that could support requirements specification and system design. Not all representatives were mindful of the need for specific cyberthreat projections, but some considered their absence to be a significant shortcoming in system development.

In the absence of threat estimates, platform designers need to postulate threats and then design to these postulated threats. The result can be that implementations of information assurance vary widely, possibly resulting in systems that are vulnerable to adversarial attack. This approach also can lead to an incoherent set of system designs when looking across the entire set of naval programs. The committee believes that cybersecurity future threat estimates are important and are needed in order to provide a complete and coordinated picture of cyberactivities that can then be factored into naval system designs.

The preceding observations lead to the following finding.

FINDING: Intelligence community projections of future cyberthreats to naval systems do not appear to exist at the level of detail needed to support development programs focusing on cyberdefense technology insertion. Such future threat projections might be difficult to develop, given the rapidly changing nature of cybertechnology, but their development and an assessment of how they might

---

[20] See Peter Finn, 2007, "Cyber Assaults on Estonia Typify a New Battle Tactic," *New York Times*, May 19, p. A01; and John Markoff, 2008, "Before the Gunfire, Cyberattacks," *New York Times*, August 13, p. A1.

[21] Jason Sherman. 2008. "DOD Draws Lessons from Cyber Attacks Against Georgia," *Inside Defense*, Washington Defense Publishers, November 10.

apply in a naval context are needed. Naval program officials who briefed the committee noted this absence and indicated the lack of future threat information to be a significant shortcoming for their program efforts. Development of the naval threat projections would require coordinated efforts across both the naval and the national intelligence communities.

Conceptual bases for characterizing physical threats to platforms are well developed and well understood. For example, an antiship missile is characterized by speed, maneuverability, radar cross section, operational tactics of employment, and so forth. In its investigations, the committee did not find an attempt to characterize cyberthreats in an analogous, conceptual way. Rather, threats are usually discussed in terms of specific examples. There appears to be no systematic taxonomy for characterizing and thinking about cyberthreats (beyond the very high level categorization of remote access, close access, and so on). This absence is one of the factors that makes future threat projections difficult to develop, as noted in the above finding.

One approach to such a taxonomy might be a "first principles" approach based on a systematic description of the points of vulnerability of generic systems. For example, to start, one recognizes that a network could be penetrated at its end hosts, intermediate nodes (e.g., routers, Domain Name Service servers), and connecting links (International Organization for Standardization layers 1 through 4). Each of those components is then decomposed further—for example, end hosts into operating systems, applications, and hardware—with each of those being decomposed further, and so on. Finally, given this vulnerabilities decomposition, one then postulates the nature of threats that could exploit the vulnerabilities. In this way one could come upon vulnerabilities that are not exploited now but could well be in the future. While the committee discussed the need for a taxonomy, based on the scope of this study it did not take steps to derive one. Organizations involved in safety assessments and trade-offs regarding operations at risk, both within the Navy and outside the Navy (for example, the National Aeronautics and Space Administration, the Federal Aviation Administration, and the Nuclear Regulatory Commission), face issues similar to those faced by the IA community. The committee suggests that new methods can be developed by starting with well-seasoned methods and modifying them to deal with the unique aspects of IA risks.[22]

Any future systems development certainly should be mindful of assessing and addressing as necessary any potential future vulnerabilities identified in this man-

[22]For example, one potential approach to addressing vulnerabilities is the countermeasure characterization (CMC) process, as described by Lubbes, which provides both the system designers and the countermeasure developer a framework process for addressing system security requirements. See Herman O. Lubbes, Network Associates, Inc., 2001, "Countermeasures Characterizations Building Blocks for Designing Secure Information Systems," IEEE 0-7695-1212-7/01, p. 103. Available at <http://ieeexplore.ieee.org/ielx5/7418/20170/00932196.pdf?arnumber=932196>. Accessed February 24, 2009.

ner. In addition, an understanding of these future vulnerabilities is necessary for guiding research and development (R&D) efforts to counter cyberthreats. R&D cannot just be directed against today's threats.

The preceding observations are summarized in the following finding.

FINDING: No systematic and widely accepted taxonomy for characterizing cyberthreats appears to exist. Such a taxonomy could be based on a first-principles characterization of the potential points of vulnerability of distributed systems. A systematic taxonomy is necessary for guiding research and development efforts and for assessing systems under development for their resilience against the whole threat spectrum.

ASSESSMENT OF CURRENT CYBER VULNERABILITIES

The vulnerability of naval and DOD systems is discussed in the context of the threat described above. This discussion is phrased in terms of trends.

Growing Use of Commercial Technology for Military Applications

The committee recognizes that the adoption of COTS technologies in the military for both mission-critical and noncritical systems is and will continue to be necessitated by economic advantages (related to economy of scale) and the advantage of speed to deployment when compared to custom-developed systems.[23] However, with the widespread adoption of COTS technologies in mission-critical networks comes the shared risk of information technology (IT)-based attacks common to COTS technologies in these networks.[24] For the military to gain both the economic and timely technological advantages of applying COTS communications and computing technologies (both hardware and software) to mission-critical systems, a corresponding set of IA risks must be taken and a corresponding set of IA strategies must be developed for managing those risks. With the adoption of COTS products, the DON also faces the added challenge of and concern with assurances regarding how their vendors treat the security of COTS products; in

[23]Additional advantages of using COTS in DOD systems include the fact that recruits are familiar with the products, which translates to potential savings and efficiencies in training.

[24]For additional background, see Samuel T. King, Joseph Tucek, Anthony Cozzie, Chris Grier, Weihang Jiang, and Yuanyuan Zhou, 2008, "Designing and Implementing Malicious Hardware," *Proceedings of the First USENIX Workshop on Large-Scale Exploits and Emergent Threats (LEET)*, San Francisco, Calif., April; "The State of Offensive Affairs in the COTS World," at <http://www.fastcompany.com/magazine/127/nexttech-fear-of-a-black-hat.html>; Brian Grow, Chi-Chu Tschang, Cliff Edwards, and Brian Burnsed, 2008, "Dangerous Fakes," *BusinessWeek*, October 2, at <http://www.caughq.org/exploits/CAU-EX-2008-0002.txt>; and SecuriTeam™, Beyond Security, 2008, *Kaminsky DNS Cache Poisoning Flaw (Exploit)*, McLean, Va., July 24. Available at <www.securiteam.com/exploits/5EPOM15OUQ.html>. All accessed February 11, 2009.

particular, shared or open knowledge about both hardware and software products can provide adversaries with insights into how to break into systems or disable them at the critical times when they are most needed. Furthermore, foreign manufacturing of products provides opportunities for the insertion of mechanisms to enable break-ins or disruptions on command.[25] In addition, the incentives of private industry to build COTS equipment are based on priorities that are different from those dictated by DOD and DON information assurance concerns.

Newer Directions in Commercial Information Technology and Naval Adoption

As computing hardware and software capabilities expand, commercial products are emerging that integrate more and more functionality into single products. Embedding user-developed application computing support into communication switches (such as the Cisco Application-Oriented Networking product line), providing for remote monitoring and control of systems (such as in Motorola's Supervisory Control and Data Acquisition systems), and adding more and more functionality into operating systems (such as Microsoft's Vista) are all examples of the trend toward greater integration. In addition, driven by immediate cost and system management advantages, COTS-based systems architectures continue to emerge that organize system administration, system management, and system service capabilities into more centrally manageable configurations. For example:

- Service-oriented architectures are permitting distributed hardware and software systems with centralized system management and administration,
- High-performance communications switches permit a single fiber-based local area network with logically controlled and isolated communications channels to replace multiple copper-based local area networks that are physically separated and have thus been administered and controlled separately, and
- The employment of automated software patching systems supporting commonly configured user machines enables automated support for rapid security patching.

A natural by-product of these trends is the adoption of more integrated commercial components into naval systems in order to gain the same advantages that commercial companies are interested in. Integration may in some cases reduce the likelihood of a successful attack; however, the potential consequences of a successful attack are greatly increased as a result of the expanded scope that the

---

[25]Defense Science Board. 2007. *Mission Impact of Foreign Influence on DOD Software,* Office of the Under Secretary of Defense for Acquisition, Technology, and Logistics, Washington, D.C., September.

attack might have as a result of the more extensive integration at the component and system levels.[26] This study observes that this extrapolation is not hypothetical and that in fact it is in progress through a variety of naval system development activities (see Chapter 4).

The discussion of this and the previous subsections is summarized in the following finding.

FINDING: The ever-growing use of commercial technology for military applications increases information assurance risks. Furthermore, the newer directions in commercial information technology (e.g., greater integration in single products) and naval adoption further exacerbate these risks.

Reactive Posture Against Cyberthreats

New cybersecurity threats and vulnerabilities are identified almost daily.[27] As new vulnerabilities emerge, new initiatives have been introduced to counter

[26]Recent articles published by the Carnegie Mellon Software Engineering Institute argue that as complexity grows, components of networked systems may sometimes process information from other systems whose intentions and trustworthiness are not always known. As a result, a hierarchical structure in a complex system has the undesirable property that every node and link of the hierarchy potentially constitutes a single point of failure for the system as a whole. That is, if the success of a function or system depends on the success of each of its components and subsystems, then an error, compromise, or failure in any one component propagates to the system as a whole and undermines system-wide success. See Carol Woody and Robert Ellison, 2007, "Survivability Challenges for Systems of Systems," Carnegie Mellon Software Engineering Institute, No. 6, Pittsburgh, Pa.; and David Fischer and Dennis Smith, 2004, "Emergent Issues in Interoperability," Carnegie Mellon Software Engineering Institute, Pittsburgh, Pa., No. 3. Both are available at <www.sei.cmu.edu/news-at-sei/columns>. Accessed February 25, 2009.

[27]For example, see CyberInsecure.com (a posting of daily cyberthreats and Internet security news alerts), May 21, 2008: "An attack, demonstrated by Rich Smith from HP Systems Security Lab at the EUSecWest security conference in London, showed that embedded systems hardware can be damaged beyond repair. The attack could be carried out remotely over the internet"; May 12, 2008: "Security researchers have discovered a new technique for developing rootkits, malicious packages used to hide the presence of malware on compromised systems. Instead of hiding a rootkit in the virtualization layer, the rootkit can be smuggled into System Management Mode (SMM), an isolated memory and execution environment supported in Intel chips that's designed to handle problems such as memory errors"; November 20, 2008: "Recent increase in malicious code propagating via USB flash drives forced the US Army to suspend the use of USB and removable media devices after a worm began spreading across its network. Use of USB drives, floppy discs, CDs, external drives, flash media cards and all other removable media devices has been placed on hold in order to contain the spread of Agent-BTZ, a variant of the SillyFDC worm"; and January, 19, 2009: "According to warnings issued by Research in Motion (RIM), hackers can use booby-trapped PDF attachments sent to BlackBerry devices to launch malicious code execution attacks. The company shipped patches this week to address a pair of critical vulnerabilities affecting their product." All accessed February 17, 2009. Weekly cybersecurity reports providing summaries and ratings of new vulnerabilities are also provided by the United States Computer Emergency Readiness Team; available at <http://www.us-cert.gov/cas/bulletins/>. Accessed February 17, 2009.

them. A common element of these initiatives is that they are reactive to the current threats; that is, there is no element focusing on possible future threats. Many of the presentations received by the committee recognized this reactive approach and expressed a desire to "get ahead" of the threat. Alternative approaches are needed to break out of this reactive mode. Despite a nearly universal desire to do so, the committee saw little evidence of efforts or a plan to develop such alternative approaches. The one significant exception is the beginning of approaches to support cyberdefense with cyber offense (see the discussion below).

The reactive posture is tied to the fact that naval IA strategy is currently based on "best commercial practices," which are largely reactive, in the sense described above. The conservative nature of the commercial marketplace has defined best practices that fall short of the security needs of the military. For example, the broad commercial marketplace for routine nonsecure applications and use will not tolerate false alarms by antivirus scanners. This has led the industry to focus primarily on signature-based detection strategies that are highly accurate at detecting already-known threats but that are blind to new threats never seen before. Basing naval forces IA strategy solely on such commercial practices will result in a reactive IA strategy for naval forces that is incapable of achieving realization of the strategic desire to get ahead of the threat.[28] Compounding this negative impact is the possibility that naval forces may face a significantly different threat from that confronting commercial industry, especially in a situation that could involve a nation-state conflict.

The above discussion is summarized in the following observation.

FINDING: Naval approaches to countering cyber vulnerabilities are primarily reactive to threats, being based largely on commercial best practices. While DON representatives who met with the committee expressed the need to "get ahead" of the threat, the committee saw little evidence that approaches to do so were being actively pursued by naval personnel.

Layered Defense Strategy for Cybersecurity

The committee observed many references to the use of the "layered defense" (or "defense-in-depth") approach to cybersecurity. In its ideal form, a layered defense has mutually supporting layers of security solutions within and among its IT assets—typically with overlapping domains so that a failure of one solution will not jeopardize the entire system—and would also include measures for both protection and detection. In actual fact, real-world controlled connection or

---

[28]The committee was briefed on cyberdefense concepts being explored at both the National Security Agency and the Defense Advanced Research Projects Agency. These emerging concepts should help the DON address the need for a more proactive strategy.

air-gap implementations[29] for cybersystems can sometimes be highly porous and subject to "end runs" by widely available technologies such as Universal Serial Bus (USB) drives and Wi-Fi connectivity.[30] Defense in depth is critical because the effectiveness of individual layers cannot be assured, but one cannot assume that each layer will "get a shot" as would happen in the defense of physical assets (e.g., strike group defense against incoming antiship missiles).

Because it is connected to the Internet, the NIPRnet introduces particular vulnerabilities to the layered-defense approach. The relatively unrestricted NIPRnet to Internet connection, exacerbated by "non-official" uses of the NIPRnet, provides an opportunity for adversaries to seek out and exploit vulnerabilities that enable elevated privileges, allowing access to inner cyberdefense layers. Even without elevated privileges, adversaries can potentially disrupt many essential functions that are carried out on the NIPRnet. Although the full set of dependencies on the NIPRnet for mission-critical military operations was not established by the committee, logistics support on the NIPRnet was identified as an important aspect of naval operations that is subject to potential compromise by an adversary. The DOD is considering tighter restrictions on the NIPRnet; however, it seems that there are mixed views across both the DOD and DON about the risks of continuing with an integrated NIPRnet, many devaluing the IA concerns relative to other, morale-related benefits of its open use.

Summary Assessment of Vulnerabilities

There is a general recognition by the Department of the Navy of the seriousness of cybersecurity vulnerabilities, as evidenced by the commission of this study. This recognition has resulted in increased attention in this area, leading to many initiatives to improve the situation. Some of these initiatives are complete and have improved the cybersecurity posture of the DON. But, naval forces are increasingly dependent on information technology systems that cannot be trusted. Mitigating the IA risks that result from this dependence will require additional approaches to supplement the reactive approach of following commercial best practices that prevails today. In the presentations that it received, the committee found little evidence of plans to develop such an alternative approach. Thus, the existing cyber vulnerabilities are expected to continue in the foreseeable future.[31]

---

[29]An air-gap defense inserts a deliberate break, to be connected by manual action, in a link of the network (see Naval Studies Board, National Research Council, 2000, *Network-Centric Naval Forces: A Transition Strategy for Enhancing Operational Capabilities,* National Academy Press, Washington, D.C., p. 36).

[30]For example, see U.S. Cyber Emergency Readiness Team, National Cyber Alert System, Cyber Security Tip ST08-001, "Using Caution with USB Devices," updated November 4, 2008. Available at <http://www.us-cert.gov/cas/tips/ST08-001.html>. Accessed February 23, 2009.

[31]The nature of the changing status of information operations and the potential impact on public and private sectors, as well as on U.S. military forces, are described in numerous reports, including unclassi-

Recognizing the plethora of possible attacks and the corresponding effort that it would take to defend against all of them, one would see that the future is defined by an attack/defense conflict that is mismatched, with the advantage heavily on the side of the attacker. In this environment, naval forces can expect that under circumstances defined by adversaries, new attacks will appear that result in the denial or disruption of network connectivity and the corruption and compromise of mission-critical data. Procedures to "fight through" such obstacles are being explored in the fleets, and the committee wishes to acknowledge these efforts and advocate their widespread development and deployment.

This assessment is summarized in the following finding.

FINDING: While valuable information assurance initiatives have been implemented, DON and DOD sources have indicated, in general, a significant deficiency in the ability to defend against the wide array of possible cyber penetration threats.

IMPORTANT FINDINGS FROM RELATED STUDIES

Several IA-related studies conducted in recent years by Federally Funded Research and Development Centers and other organizations were discussed with the committee.[32] A summary of these studies is included in Appendix D of the present report. In addition, the committee was briefed in depth on two important IA-related advisory board studies (see the subsections below). The committee found that the major themes derived by each of the studies, when taken together, should form an important part of the basis for the Department of the Navy's development of a strategy for addressing its future IA needs.

Air Force Scientific Advisory Board Study

The key findings of a 2007 study by the Air Force Scientific Advisory Board (AFSAB)[33] on the implications of cyberwarfare are the following:

---

fied reports to Congress. For example, see U.S. Government Accountability Office, 2007, *Cyber Crime: Public and Private Entities Face Challenges in Addressing Cyber Threats,* Report to Congressional Requesters, Washington, D.C., June; John Rollins and Clay Wilson, 2007, *Terrorist Capabilities for Cyber Attack: Overview and Policy Issues,* Congressional Research Service, Washington, D.C., January 22; and U.S. Government Accountability Office, 2008, *Cyber Analysis and Warning: DHS Faces Challenges in Establishing a Comprehensive National Capability,* GAO-08-588, Report to Subcommittee on Emerging Threats, Cybersecurity, and Science and Technology, Committee on Homeland Security, House of Representatives (Table 2, p. 7, Sources of CyberThreats), Washington, D.C., July.

[32]Michael McBeth, Office of Naval Research Advisor, and Lawrence Lynn, Center for Naval Analyses Representative, "Current Naval Research Information Assurance Studies," presentation to the committee, April 28, 2008, Naval Network Warfare Command, Norfolk, Va.

[33]Thomas F. Saunders, Chair, USAF Scientific Advisory Board Summer Study, "Implications of Cyber Warfare," presentation to the committee, March 6, 2008, Washington, D.C.

- Forces are not prepared to fight through a sophisticated, covert cyber-attack; and
- Commercial technology is not going to provide a solution for such attacks.

The AFSAB emphasized that vulnerabilities exploited by sophisticated cyber-attacks are inevitable. Thus, the Air Force needs to be prepared with technologies and with operating concepts and procedures to "work through" such attacks. The findings of this committee are consistent with those of the AFSAB.

Defense Science Board Study

According to the Defense Science Board (DSB) study chairs, the findings and recommendations of the DSB study on information management for network-centric operations published in 2007 can be distilled to three points:[34]

- The combat information capability must be treated as a critical defense weapon system.
- Information assurance must be resourced and its risk managed accordingly.
- An innovative acquisition strategy is required to leverage commercial off-the-shelf information technology while managing the IA risks.

Like the AFSAB, the DSB believes that the "system and its capabilities will always be under attack and, as a result, will always be operated in either a degraded or compromised mode."[35] Given this belief and the DSB's first finding, IA becomes a critical warfighting need, not just a support function. The DSB notes that information assurance enables mission assurance, and states that a formal risk management process is needed to assess the benefits of the added applications against the impact of the introduced information assurance threats.

The implementation status of recommendations from these reports is at various stages. However, many aspects of information assurance and related cyberwarfare operations are currently undergoing comprehensive reviews and policy updates by the DOD and each of the military services.

---

[34]Defense Science Board. 2007. *Defense Science Board 2006 Summer Study on Information Management for Net-Centric Operations,* Office of the Under Secretary of Defense for Acquisition, Technology, and Logistics, Washington, D.C., April, p. 7.

[35]Defense Science Board. 2007. *Defense Science Board 2006 Summer Study on Information Management for Net-Centric Operations,* Office of the Under Secretary of Defense for Acquisition, Technology, and Logistics, Washington, D.C., April, p. 88.

2

The Immediate Response—Current Information Assurance and Cyberdefense Initiatives

Information assurance (IA) is defined in Department of Defense (DOD) instruction documents as "measures that protect and defend information and information systems by ensuring their availability, integrity, authentication, confidentiality, and non-repudiation. This includes providing for restoration of information systems by incorporating protection, detection, and reaction capabilities."[1]

Additionally, the DOD's long-term vision for an effective network-centric operating environment—and the associated Global Information Grid (GIG)[2]—present a vision for DOD information assurance capabilities and practices that provide the following:

- Transactional Information Protection—granular end-to-end security controls that enable protected information exchanges within the variable-trust net-centric environment;

[1]Department of Defense. 2003. Department of Defense Instruction 8500.2. Information Assurance Implementation, Washington, D.C., February.

[2]As defined in DOD Instruction 8500.2, ibid., the GIG consists of the "globally interconnected, end-to-end set of information capabilities, associated processes, and personnel for collecting, processing, storing, disseminating, and managing information on demand to war fighters, policy makers, and support personnel." Included are all government-owned and -leased communications and computing systems and services, as well all software, data, security services, and anything else necessary to operate and secure the GIG. Also included are the National Security Systems as defined in Section 5142 of the Clinger-Cohen Act of 1996 (National Defense Authorization Act for FY 1996, Public Law 104-106, formerly called the "Information Technology Management Reform Act," February 10, 1996). By this definition, the GIG encompasses all DOD and National Security information systems at all levels, from tactical to strategic, as well as the interconnecting communications systems.

- Digital Policy-Enabled Enterprise—dynamic response to changing mission needs, attacks, and system degradations through highly automated and coordinated distribution and enforcement of digital policies;
- Defense Against an Adversary From Within—persistently monitor, detect, search for, track, and respond to insider activity and misuse within the enterprise;
- Integrated Security Management—dynamic automated net-centric security management seamlessly integrated with operations management; and
- Enhanced Integrity and Trust of Net-Centric Systems—robust IA embedded within enterprise components and maintained over their life cycle.[3]

Because of the interconnected nature of the GIG, IA is a shared need and capability across the DOD and its Services. Each of the Services is responsible for the development of its own network-related mission and structures, and also for the control and defense of information on its portion of the GIG. Because naval nodes of the GIG are integrated with non-naval nodes, a gap in one area of GIG information assurance capability has the potential to impact other areas.

It is broadly recognized, however, that the GIG IA vision stated above is not a current reality; therefore, the Department of the Navy (DON), as well as the DOD and other Services, has information assurance and cyberdefense initiatives underway to improve protection against the current threats to its networks and to help bring the network-centric enterprise closer to the stated IA vision. Owing to GIG interconnectivity, IA initiatives across broader sectors of the DOD are also very important to the Navy. (See Chapter 6 for a description of DOD, Navy, and Marine Corps network defense responsibilities.)

During the course of its data gathering, the committee was briefed on both naval and DOD-wide IA-related initiatives currently underway, and by all of the obvious organizations with direct and indirect information assurance responsibilities that might impact naval forces.[4] However, in spite of the fact that the achievement of greater information assurance requires the integration of a number of contributing solutions, no one party was able to present the committee with a comprehensive list of naval or DOD-wide initiatives. Rather, each party primarily focused on the initiatives under its individual purview.[5] In addition to receiving these presentations, the committee also performed independent research to gain more understanding of the initiatives. The following sections present a summary

---

[3]Department of Defense Chief Information Officer. 2007. *Global Information Grid Architectural Vision: Vision for a Net-Centric, Service-Oriented DOD Enterprise,* Version 1.0, Department of Defense, Washington, D.C., June, p. 24. Available at <http://www.defenselink.mil/cio-nii/docs/GIGArchVision.pdf>. Accessed November 17, 2008.

[4]A description of the committee's data-gathering sessions is provided in the Preface of this report.

[5]The committee was briefed by the portfolio manager of the GIG Information Assurance Portfolio program, which is developing such a comprehensive listing (Richard Scalco, GIG IA Portfolio Manager, "GIG IA Portfolio Management Office," presentation to the committee, July 16, 2008, National Security Agency, Fort Meade, Md.).

and discussion of these initiatives, organized according to the major sources of input.

DEPARTMENT OF THE NAVY CHIEF INFORMATION OFFICER INFORMATION ASSURANCE INITIATIVES

A presentation to the committee from the office of the Department of the Navy, Chief Information Officer (DON CIO), showed the Department of the Navy to be positively engaged with the DOD in its planning and execution of DOD-wide IA initiatives.[6] This relationship appears to be fruitful in that it provides the DON with the ability to leverage DOD capabilities that reside outside the DON (such as the digital signature and encryption capabilities provided by the DOD to help verify user identity). In addition to leveraging DOD-wide capabilities for naval forces, activities such as the promulgation of the Department of the Navy's vision and plans for its Next Generation Enterprise Network (NGEN) and the implementation of the Navy's Prometheus[7] system for monitoring and analyzing network information are positive steps for improving IA across the Navy's Network Warfare (NETWAR)/FORCEnet Enterprise.

The DON Deputy Chief Information Officer (DCIO) serves as the Senior Information Assurance Officer (SIAO) of the Department of the Navy. One particularly interesting initiative that the DCIO briefed to the committee was the effort to establish a cyber task force involving the DON CIO, the Office of the Chief of Naval Operations, the United States Marine Corps, and the Naval Criminal Investigative Service.[8] This task force would be chaired by the DON SIAO and have oversight from the Deputy Under Secretary of the Navy/DON CIO. The objectives of the task force would be as follows:

- To articulate the process for coordinating computer network attack (CNA), computer network exploitation (CNE), and computer network defense (CND) with the DON, as well as counterintelligence (CI) for these activities;
- To ensure feedback from CNA and CNE activities into CND planning and execution, and to ensure that a similar feedback loop exists for CI activities;
- To provide a complete and coordinated picture of cyber activities within the DON;
- To ensure a synchronized and coordinated investment in cyber activities; and

[6]John Lussier, Department of the Navy Deputy Chief Information Officer, "Department of the Navy CIO Organization," presentation to the committee, March 6, 2008, Washington, D.C.

[7]Prometheus is the name given to an information technology system recently implemented by the Navy Cyber Defense Operations Command (NCDOC) to provide both network protection and network situational awareness for NCDOC-defended networks.

[8]John Lussier, Department of the Navy Deputy Chief Information Officer, "Department of the Navy CIO Organization," presentation to the committee, March 6, 2008, Washington, D.C.

- To align roles and responsibilities so as to enable timely execution of cyber-related policies, to aid in the implementation of cyber products, to provide a well-defined governance of cyber practice, and to have a focused, coordinated cyber investment practice.

At the time of the briefing by the DCIO (March 2008), establishment of the task force was pending approval by the Secretary of the Navy. If established, the task force could address significant issues, such as the coupling of CND, CNA, and CNE.[9]

The DCIO provided the committee with a list of other DON IA initiatives, presented in Table 2.1. The DCIO also provided the committee with a list of DOD-wide IA initiatives that are being addressed by naval forces. (The DOD-wide IA initiatives reported by the DON CIO are included in Table 2.2 and are discussed in the subsection on "Defense-Wide IA Initiatives.") The list of DON initiatives presented in Table 2.1 is not complete, as can be seen by comparing it with the naval initiatives discussed in the subsections below addressing IA initiatives sponsored by the Naval Network Warfare Command (NETWARCOM), the Navy Information Systems Security Program (ISSP), the Navy's Space and Naval Warfare Systems (SPAWAR), and other fleet forces operations.

NAVAL NETWORK WARFARE COMMAND INFORMATION ASSURANCE INITIATIVES

The Naval Network Warfare Command has two major responsibilities: It functions as (1) a type commander[10] and (2) an operational commander. In the former role its responsibility is to organize, train, and equip for network operation, just as other type commanders do in their respective areas. However, NETWARCOM is not directly involved in acquisition. In its latter role, NETWARCOM manages networks and network security, ranging from the Navy/Marine Corps Intranet (NMCI) down to the Network Operations Center level. In addition to their other discussions with the committee concerning IA issues and policies, NETWARCOM personnel presented the following as NETWARCOM's major IA initiatives, several of which are also being implemented by the Marine Corps and the Marine Corps Network Operations and Security Command (MCNOSC):[11]

[9]For example, see Maj Donald W. Cloud, Jr., USAF, 2007, "Integrated Cyber Defenses: Towards Cyber Defense Doctrine," Master of Arts Thesis, Naval Postgraduate School, Monterey, Calif., December. Available at <https://www.hsdl.org/homesec/docs/theses/07Dec_Cloud.pdf&code=a469b8967301e4226f41c61fcc2706b3>. Accessed February 26, 2009.

[10]In the U.S. Navy, the type commander is the flag officer responsible for all ships of a certain type in the fleet.

[11]Alan L Rickman, Naval Network Warfare Command, "Decision Superiority for the Warfighter," presentation to the committee, March 5, 2008, Washington, D.C.

TABLE 2.1 Department of the Navy Current Information Assurance Initiatives: Selected List

Fiscal Year (FY) of First Impact			
FY 2008	FY 2009	FY 2010	FY 2011 or Beyond
Cryptographic Log-on Policy Enforcement Tools for Access (research) Attribute-Based Access Control (pilot) Secretary of the Navy Warning Orders Wireless Security Cyber Asset Reduction and Security	Data at Rest Encryption Navy/Marine Corps Intranet "Sweet 16"[a]	Thin-Client coupled with Virtual Machine concept Next Generation Enterprise Network Security Plan and Concept of Operation	Next Generation Enterprise Network

[a]The Navy/Marine Corps Intranet information assurance initiatives—commonly referred to as the "Sweet 16"—are discussed in the subsection entitled "Navy/Marine Corps Intranet" in the present chapter and are presented in Table 2.4.
SOURCE: Derived from information presented to the committee by John Lussier, Department of the Navy Deputy Chief Information Officer, "Department of the Navy CIO Organization," March 6, 2008, Washington, D.C.

- *Operational Designated Approval Authority.* Provides an end-to-end approach for certification and accreditation (C&A) processes. This initiative is targeted at reducing C&A cycle time.
- *Public Key Infrastructure (PKI).* Implements the requirement for cryptographic network log-on across all naval unclassified networks. Eliminates "stovepipe" solutions and requires the use of common access cards (CACs) across protected naval systems. Also includes the investigation of biometrics identification.
- *IA Computer Network Defense.* Provides monitoring of all naval networks, analyzes trends, and develops mitigating strategies. Regularly reviews all policy and procedures and provides security relationship with Navy industrial base and contractor networks.
- *Data at Rest.* Provides encryption for all mobile computing devices and removable media processing for controlled unclassified information and Personal Identifiable Information.[12]

[12]Personal Identifiable Information, or PII, is defined by an Office of Management and Budget memorandum (Karen S. Evans, Administrator, Office of E-Government and Information Technology,

TABLE 2.2 Department of Defense-Wide Current Information Assurance Initiatives: Selected List

Fiscal Year (FY) of First Impact			
FY 2008	FY 2009	FY 2010	FY 2011 or Beyond
Department of Defense Demilitarized Zone[a] Certification and Accreditation Public Key Infrastructure (ongoing) Enterprise Standards Across Common Architecture (ongoing) Trusted Computing Consortium (ongoing) Joint Task Force–Global Network Operations Security Awareness Messages	Department of Defense Training Initiative (ongoing)	Non-Classified Internet Protocol Router Network (NIPRnet) Deep Dive (Controlled Unclassified Information Behind Demilitarized Zone After Deep Dive) Supply Chain Risk Management	Global Information Grid Mission Assurance Plan

[a]The Demilitarized Zone, or DMZ, approach to defending the Global Information Grid provides a separate interface to the Internet and external DOD connections, thus limiting vulnerabilities to malicious attacks, worms, and viruses that plague the Internet.
SOURCE: Derived from information presented to the committee by John Lussier, Department of the Navy Deputy Chief Information Officer, "Department of the Navy CIO Organization," March 6, 2008, Washington, D.C.

- *Cyber Asset Reduction and Security.* Reduces the number of legacy networks and hence reduces the vulnerabilities inherent in those networks.
- *Wireless Security.* Provides technology guidance for wireless solutions and guidance for the resulting expanded mobility of the GIG. This initiative also includes Secure Blackberry, that is, CAC-based PKI to sign and encrypt wireless e-mail.

Executive Office of the President, OMB Memorandum for Chief Information Officers, M-06-19, Washington, D.C., July 12, 2006) as "information which can be used to distinguish or trace an individual's identity such as their name, social security number, date and place of birth, biometrics records, etc., alone, or when combined with other personal or identifying information which is linked or linkable to a specific individual, such as laptop computers, thumb drives and personal digital assistants (PDAs)."

Many of these NETWARCOM information assurance initiatives are also reflected in the DON information assurance initiatives presented in Table 2.1.

INFORMATION SYSTEMS SECURITY PROGRAM INITIATIVES

The Information Systems Security Program is the Department of the Navy's research, development, testing and evaluation (RDT&E) program element, which includes the DON's individual information assurance projects.[13] The program is projected to be funded at approximately $30 million per year for the period fiscal year (FY) 2008 through FY 2013 and includes both research and development (R&D) and technology implementation funds. The Navy's ISSP is described as follows in the document prepared for FY 2009 budget justification:

> The Navy ISSP RDT&E program works to provide the Navy with these essential Information Assurance elements: (1) Assured separation of information levels and user communities, including coalition partners; (2) Assurance of the telecommunications infrastructure; (3) Assurance of Joint user enclaves, using a defense-in-depth architecture; (4) Assurance of the computing base and information store; and (5) Supporting assurance technologies, including a Public Key Infrastructure (PKI) and directories. The goal of all ISSP RDT&E activities is to produce the best USN operational system that can meet the certification and accreditation requirements outlined in DoD Instruction 5200.40 (new DoDI 85xx series pending). Modeling DoD and commercial information and telecommunications systems evolution (rather than being one-time developments), the ISSP RDT&E program must be predictive, adaptive, and technology coupled. The program develops frameworks, architectures, and products based on mission threats, information criticality, exploitation risks, risk management, and integrated Joint information system efforts.[14]

The key ISSP projects listed in the Navy Exhibit R-2 RDT&E program are summarized and described in Table 2.3. The single largest individual FY 2009 budgetary item in the Navy's ISSP is the Navy Cryptographic Modernization Program and its associated secure communications, budgeted at $8.75 million in this particular program element.[15]

---

[13]The ISSP effort is a naval enterprise-wide responsibility derived from requirements outlined in the Secretary of the Navy Instruction 5239.3A, Department of the Navy Information Assurance Policy (INFOSEC) Program, Washington, D.C., December 20, 2004.

[14]Department of the Navy. 2008. "Department of the Navy Exhibit R-2 RDT&E Budget Item Justification," Washington, D.C., February, p. 2.

[15]The budget numbers in this R&D exhibit reflect only a limited portion of the total budget for the Navy Cryptographic Modernization program. The $8.75 million is referred to here only to show its size in relation to the $30 million RDT&E total for this ISSP program element mentioned earlier.

TABLE 2.3 Information Assurance (IA) Initiatives in the Navy's Information Systems Security Program

IA Project Name	Project Description
Computer Network Defense	Develops and implements an integrated system of filters, firewalls, intrusion prevention systems, patch management, encryption, and other vulnerability remediation tools and policies for fleet and ashore networks.
Cryptographic Modernization	In coordination with Joint Services and the National Security Agency, provides development support, specifications, acquisition documentation, and testing for identified and selected cryptographic products to provide secure communications. Replaces decertified systems in accordance with Joint Chiefs of Staff modernization schedule.
IA Readiness	Provides systems security engineering support to all Department of the Navy organizations in the certification and accreditation of information systems.
Secure Voice	Completes the development and integration test of the Secure Communication Interoperability Protocol Inter-working Function for off-ship secure communication capabilities while underway.
Cross Domain Solutions	Provides system security engineering development, testing, and evaluation for multilevel security solutions (databases, Web browsers, routers/switches, etc.), for allied and coalition participation.
Key Management Infrastructure	Develops advanced key management security testing, certification, and accreditation for various naval systems.
Emerging Technologies	Supports the development of Department of the Navy information assurance architectures and the transition of new technologies addressing Navy information assurance challenges.

SOURCE: Department of the Navy. 2008. "Department of the Navy Exhibit R-2 RDT&E Budget Item Justification," Washington, D.C., February, p. 2.

INFORMATION TECHNOLOGY AND NETWORK PROGRAMS INFORMATION ASSURANCE INITIATIVES

Much of the Navy's information assurance activity is embedded in information technology (IT) and network programs associated with large specific naval program activities, in addition to the targeted IA-focused projects of the ISSP. The committee was briefed in detail on three such major programs: the Navy/Marine Corps Intranet, the planned Next Generation Enterprise Network (a follow-on to NMCI), and the Navy's Consolidated Afloat Networks and Enterprise Services (CANES). The information assurance components of these major programs, as highlighted for the committee, are summarized below.

Navy/Marine Corps Intranet

The Navy/Marine Corps Intranet, with more than 650,000 users, is reported to be the largest corporate intranet in the world, and also to represent the single largest government IT contract.[16] Although NMCI is currently managed through a contracted outsource organization, the Navy's NETWARCOM, through its Global Network Operations Center, provides IA and network defense oversight for the Navy enclave of NMCI, and MCNOSC provides IA and network defense oversight for NMCI's Marine Corps enclave. Thus, while NMCI daily operations are managed externally, many of the current DON and DOD IA initiatives are being applied, where appropriate, to the NMCI system. A list of the top 16 current NMCI network security initiatives is provided in Table 2.4; all are scheduled to be implemented before NMCI transitions to NGEN in 2010.

Next-Generation Enterprise Network

Current plans are for the Next Generation Enterprise Network to encompass the current Navy/Marine Corps Intranet, plus the Overseas Navy Enterprise Network (ONE-Net), the remaining "legacy" networks, the Navy's shipboard IT for the 21st Century (IT-21) networks, and the Marine Corps Enterprise Network (MCEN).[17] Thus, many of the security features that have recently been added to NMCI will likely be integrated from the beginning and enhanced for NGEN. (See Figure 2.1 for a visual diagram of the relationships among these currently existing naval network systems.)

As reported to the committee, it is anticipated that future NGEN upgrades will transition NMCI, ONE-Net, IT-21, and MCEN from four separately managed environments to a globally integrated, network-centric DON enterprise to support network operations (NETOPS) and leverage the DOD Global Information Grid and available DOD enterprise services. This integration effort promises to improve information assurance for the largest network across the entire naval network-centric enterprise. Also, a key IA advancement of NGEN over NMCI is

---

[16]Terrelle C. Bradshaw, Naval Network Warfare Command, Global Network Operations Center, "NMCI IA Overview," presentation to the committee, April 29, 2008, Norfolk, Va. NMCI also is reported to support more than 100 million e-mail messages per month and 124 million browser transactions per day, and to provide connectivity for approximately 11,000 wireless communication devices. The running of NMCI daily operations is contracted to Electronic Data Systems in a 10 year contract, extending from October 2000 to October 2010.

[17]RADM(S) David G. Simpson, USN, Director, Navy Networks, Deputy Chief of Naval Operations, Communication Networks (N6), "Next Generation Enterprise Network (NGEN) and Consolidated Afloat Networks and Enterprise Services (CANES)," presentation to the committee, May 29, 2008, Washington, D.C. The planned baseline for NGEN is 340,000 workstations; approximately 650,000 user accounts; support for mobile devices; and the associated network operations command and control. NGEN is currently scheduled to phase into operation at the end of the NMCI contract, which expires in October 2010.

TABLE 2.4 Current Information Assurance (IA) Initiatives for the Navy/Marine Corps Intranet

Initiative	Description
Intrusion Protection System	Upgrades intrusion detection infrastructure.
Logging Infrastructure	Integrates logging infrastructure to support network audits and incident response.
Firewall Suites	Implements improved firewall protection.
Improved Public Key Infrastructure (PKI)	Implements Service-wide e-mail signing and encryption.
Improved IA Vulnerability Alert Management	Improves reliability of IA vulnerability patching and implements Network Access Control.
Host Based Security System	Implements the DOD enterprise-wide automated and standardized tool to provide end-point (server, desktop, and laptop) security against both insider threats and external threats that are able to penetrate boundary defenses. Provides centralized management of host-based capabilities.
Data At Rest Encryption	Provides encryption for all mobile computing devices and removable media.
Network Configuration Management	Provides and maintains current network configuration data and assures continuous access for security testing and evaluation.
Two Factor Authentication	Enables system administrator to provide improved authentication for all accounts.
PKI for Blackberry	Provides PKI support for Blackberry e-mail.
Network Forensics	Establishes a network-based forensics tool for imaging system hard drives involved in an IA incident.
Security Event Management	Implements system to provide security information management compatible with other Navy and Marine Corps systems.
Common Access Card Support	Provides Web access authenticated by the common access card.
Secure Configuration Compliance Validation Initiative (SCCVI)/Secure Configuration Remediation Initiative (SCRI)	Implements DOD-recommended tools to discover assets and identify known security vulnerabilities (SCCVI), and implements corrective actions to mitigate a vulnerability (SCRI).
Uniform Resource Locator/ Content Filtering	Provides advanced-application firewall technology to update and replace aging, existing system application.
Global Access List	Updates access directories and provides certificates allowing synchronization across military components.

SOURCE: Derived from information presented to the committee by Terrelle C. Bradshaw, Naval Network Warfare Command, Global Network Operations Center, "NMCI IA Overview," April 29, 2008, Norfolk, Va.

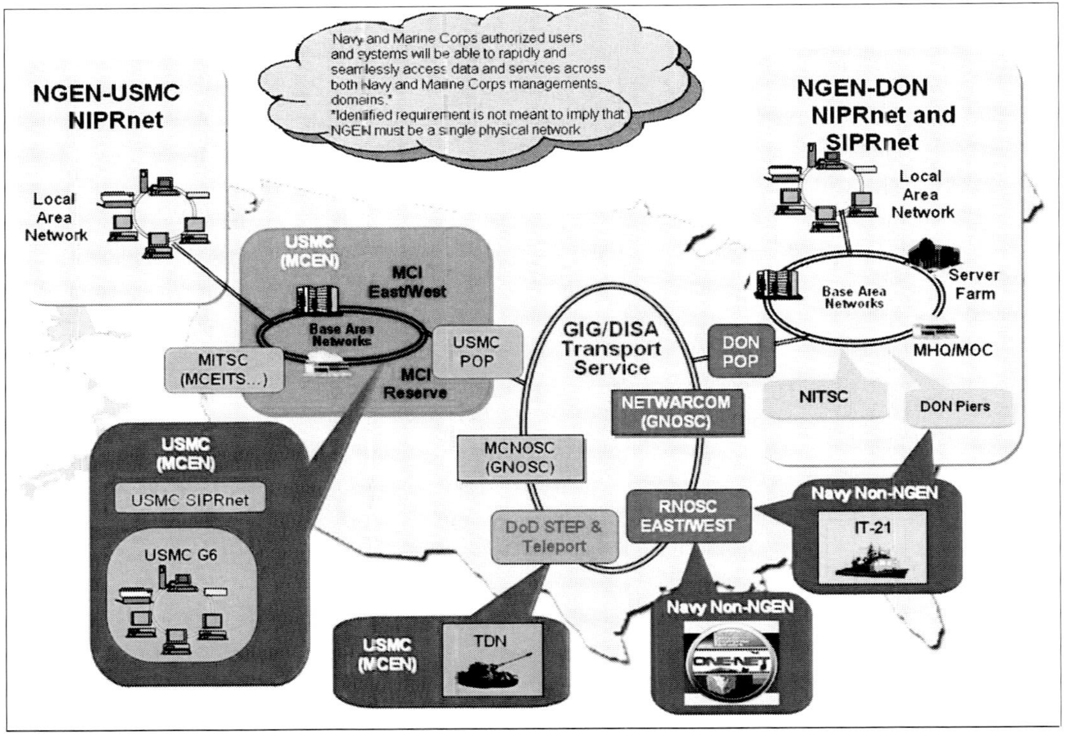

FIGURE 2.1 System relations for the Next Generation Enterprise Network (NGEN). NOTE: Acronyms are defined in Appendix A. SOURCE: RADM(S) David G. Simpson, USN, Director, Navy Networks, Deputy Chief of Naval Operations, Communication Networks (N6), “Next Generation Enterprise Network (NGEN) and Consolidated Afloat Networks and Enterprise Services (CANES),” presentation to the committee, May 29, 2008, Washington, D.C.

postulated by NGEN program management to be the inherently improved security governance for NGEN, as DON will have full visibility into the network. This will likely be the case if IA and network defense for NGEN are managed in-house, as is currently planned according to public reports, rather than managed through a contract organization as is the current situation with NMCI.

Consolidated Afloat Networks and Enterprise Services

The Navy's Consolidated Afloat Networks and Enterprise Services program is primarily a system redesign and acquisition program for afloat networks; however, it can also be viewed as a broad initiative designed to consolidate and reduce network infrastructure,[18] reduce legacy systems aboard ships, and provide increased network capability to the afloat platform enclaves. Key IA-related initiatives included in the CANES common computing environment are its built-in computer network defense capabilities, its cross-domain solutions, and its utilization of service-oriented architectures (SOAs). The committee recognizes both advantages and disadvantages of such a network architecture approach,[19] and thus it devotes additional dedicated sections to IA architecture and SOAs in Chapter 4.

The CANES program time line, as reviewed with the committee, is multiyear, with a planned 2008-2016 implementation. However, CANES "early adopters," beginning in 2009, will be permitted to test the program's key IA architectural features, allowing an opportunity to adapt the required SOA IA program features and to address requirements for hardening the NIPRnet architecture. The committee views CANES early adopters as providing an opportunity to establish an important testbed for IA advancements for security afloat, with great potential leverage.

[18]Today there are four primary shipboard infrastructure networks—NIPRnet, SIPRnet, the Joint Worldwide Intelligence Communications System, and the Combined Enterprise Regional Information Exchange System—each operating at different security levels.

[19]Many of the existing solutions to IA problems (and many of the requirements in existing IA regulations) assume that both clients and servers are located on the same physical or logical network. The clients and servers rely heavily on perimeter or boundary protection such as demilitarized zones, firewalls, and intrusion detection to prevent security threats. However, the interoperability and loose coupling requirements of an SOA necessitate additional security capabilities to complement those security models. For example, see the report on Net-Centric Enterprise Solutions for Interoperability, a collaborative activity of the U.S. Navy Program Executive Office for Command, Control, Communications, Computers and Intelligence and Space, the USAF Electronic Systems Center, and the Defense Information Systems Agency, 2006, *Net-Centric Implementation Framework*, V1.3, June 16. Available at <http://nesipublic.spawar.navy.mil/docs/part2/NESI_Part_2_v1pt3pt0-16Jun06.pdf>. Accessed November 19, 2008.

SPACE AND NAVAL WARFARE SYSTEMS COMMAND AND PEO C4I INFORMATION ASSURANCE INITIATIVES

Naval system commands must be aware of and respond to IA initiatives and architecting requirements as dictated by Navy and/or DOD instructions. Thus the committee was briefed by the Navy's engineering command at SPAWAR and its Program Executive Office for Command, Control, Communications, Computers and Intelligence (PEO C4I) personnel for discussion of key information assurance-related initiatives and associated issues.

SPAWAR/PEO C4I personnel are responsible for IA architecture for their domain of responsibility, ranging from the ashore network operating centers to ships afloat, and are working to build out "defense in depth" for that domain. As such, initiatives are also being led by SPAWAR personnel to cope with potential degradations due to attacks on various components of the information architecture. For the purposes of this activity, the IA competencies drawn upon at SPAWAR evolve from their expertise in information system security engineering. Such system security engineering concepts have been applied to the technologies associated with a host of development systems, including CANES, satellite communication platforms, and the Joint Tactical Radio Systems. Major IA initiatives reported to the committee by SPAWAR's information assurance organization are reflected in those initiatives previously discussed and reported in Tables 2.1, 2.2, and 2.3. For the purpose of brevity, these initiatives are not listed separately in this report. However, in addition to the previously reported naval initiatives, SPAWAR's PEO C4I and PEO Space personnel are also the primary responsible Navy party for designing and engineering system-wide defense-in-depth concepts; they are also the responsible party for developing IA architecture guidance as it relates to the execution of SOA implementation in naval systems.[20]

FLEET INFORMATION ASSURANCE INITIATIVES

The committee held discussions with the Commander, U.S. Pacific Fleet, and Pacific Fleet senior technical advisers; senior staff representatives, U.S. Third Fleet; and with command and network personnel on the USS *Normandy* (CG-60), to better understand the impact of information assurance on fleet operations and fleet missions. Several initiatives are currently underway that should be beneficial at the fleet level for operating through cyberattacks and degraded network capabilities. Specifically, the committee believes that the cyber-defense-related work underway in the Pacific Fleet, and its associated engineering developments

---

[20]For example, see the report on Net-centric Enterprise Solutions for Interoperability, a collaborative activity of the U.S. Navy PEO C4I and Space, the USAF Electronic Systems Center, and the Defense Information Systems Agency, 2006, *Net-Centric Implementation Framework*, V1.3, June 16. Available at <http://nesipublic.spawar.navy.mil/docs/part2/NESI_Part_2_v1pt3pt0-16Jun06.pdf>. Accessed November 19, 2008.

in SPAWAR/PEO C4I, should be strongly supported and adopted more broadly across naval forces as these current initiatives prove themselves.

DEPARTMENT OF DEFENSE-WIDE INFORMATION ASSURANCE INITIATIVES

The DON CIO provided the committee with a list of DOD-wide IA initiatives currently impacting naval forces (see Table 2.2 for the list, organized by year of major impact for naval forces).

In addition to these activities, the committee was also briefed on DOD information assurance and current DOD-sponsored IA initiatives from the Office of the Deputy Assistant Secretary of Defense for Information and Identity Assurance (ODASD[I&IA]). IA initiatives information received from the ODASD(I&IA) is summarized in Table 2.5. As of the writing of this report, the ODASD(I&IA) is preparing a more comprehensive strategic approach to IA initiatives.

Related to but separate from this effort is a DOD-wide GIG IA portfolio management program—the GIG Information Assurance Portfolio, or GIAP—currently being undertaken to help analyze and give input to DOD and military Services regarding strategic IA investments. While the GIAP is organized under the Office of the Assistant Secretary of Defense, Networks and Information Integration, its management is currently headquartered at the National Security Agency (NSA), the designated lead agency for defining DOD GIG IA architecture. The GIAP uses a set of broad strategic categories to track IA initiatives slightly different from the categories provided to the committee by ODASD(I&IA). Also, in this program, the GIAP claims responsibility for leading the "enterprise enabling" IA initiatives, such as the Public Key Infrastructure and the Key Management Infrastructure, across the DOD. The list of IA initiatives presented to the committee from the GIG IA portfolio viewpoint is contained in Table 2.6.

The committee also received information assurance briefings from the U.S. Strategic Command's Joint Task Force–Global Network Operations (JTF–GNO) and from the Defense Information Systems Agency (DISA). JTF–GNO directs the operation and defense of the Global Information Grid in support of DOD's full spectrum of missions.[21] DISA serves as a DOD enterprise-wide organization with an agenda to help provide information assurance tools and services in support of DOD network-centric operations. DISA has responsibility for coordinating with other federal agencies and industry to provide security configuration guides, checklists, scanning tools, and other standards to properly configure and manage applications, devices, and enclaves across the GIG for U.S. military command. DISA also plans for, acquires, and deploys enterprise-wide tools and capabilities that improve defense, attack sensing and reaction, and situational awareness. In its

[21] For additional information, see JTF–GNO fact sheet at <http://www.stratcom.mil/factsheets/gno.html>. Accessed October 21, 2008.

TABLE 2.5 Office of the Deputy Assistant Secretary of Defense for Information and Identity Assurance: Summary of Information Assurance (IA) Initiatives

IA Strategic Area	Example IA Initiatives
Protecting Core Networks	Demilitarized Zone, Firewalls, Network Sensors
Network Resiliency	Architecture for Resilience
Assured Information Access	Privileged Management
IA Systems/Platforms	IA Acquisition
Cyber Operations	Computer Emergency Response Teams
Cross Domain Sharing	Coalition Forces Interoperability and Assurance
Globalization/Supplier Assurance	Supply Chain Risk Management, Software and Hardware Assurance
Defense Industrial Base	Vulnerability Reporting Process
Identity Assurance	Public Key Infrastructure Deployment
Research Technology Insertion	Defense Advanced Research Projects Agency and IA Research
Training/Education IA/Personnel Readiness	Workforce Certification
International Readiness	International IA Best Practices
Cryptographic Modernization	High Assurance Internet Protocol Encryptor
Key Management	Key Management Infrastructure

SOURCE: Derived from information presented to the committee by Robert Lentz, Deputy Assistant Secretary of Defense for Information and Identity Assurance, "Overview of Department of Defense IA-Related Responsibilities, Initiatives, Strategies, and Studies," Washington, D.C., March 5, 2008.

TABLE 2.6 DOD and DON Information Assurance (IA) Initiatives from the GIG IA Portfolio Perspective

IA Strategic Area	Example IA Initiatives
Confidentiality (Protect Data and Networks)	Cryptographic Modernization, High Assurance Internet Protocol Encryptor, Secure Voice, Edge Systems
Computer Network Defense (Defend the Global Information Grid [GIG])	Demilitarized Zone, Host-Based Security Systems
Assured Information Sharing	Cross Domain Sharing, Multinational Information Sharing
Enterprise Security Management	Key Management Infrastructure, Pubic Key Infrastructure, Privileged Management
Foundational	IA Training, Enterprise-Wide Certification and Accreditation, Best Practices

SOURCE: Derived from information presented to the committee by Richard Scalco, GIG IA Portfolio Manager, "GIG IA Portfolio Management Office," July 16, 2008, Fort Meade, Md.

strategic document, DISA reports several key IA initiatives underway, including IA-related initiatives to accomplish the following:[22]

- Provide standard coalition information-sharing capabilities;
- Deploy cyber identity credentials throughout the GIG for safer and broader sharing;
- Continually assess the Public Key Infrastructure architecture for effectiveness;
- Redesign the NIPRnet and SIPRnet, including certain shared components (e.g., the Domain Name System), to dramatically enhance security and improve sharing;
- Develop and operate strengthened gateways between DOD and the Internet and between DOD, other U.S. networks, and coalition networks; and
- With the Services and agencies, plan and execute the movement of all publicly visible and partner-facing applications and services into demilitarized zones to improve sharing and security.

Based on the committee's collective inquiries, it appears that a single, comprehensive view of IA initiatives across the DOD does not exist. Although none of the groups referred to above provided the committee with a single, comprehensive view of DOD-wide IA initiatives, the committee constructed its own comprehensive view by piecing together the information received from these separate sources. As solutions start to move into the application layer of DOD information systems, the gaining of a comprehensive view will become more difficult, because these solutions might then reside with individual enclave managers. It will require significant efforts to achieve this comprehensive view—something that the committee views as necessary in order to select and synchronize integrated IA solutions.

OTHER INFORMATION ASSURANCE INITIATIVES

In addition to previously discussed information assurance initiatives, the committee was also briefed on work underway at the Defense Advanced Research Projects Agency and at NSA, and it received an overview of research currently included in the Comprehensive National Cyber Security Initiative. Although the details of work in these three areas cannot be discussed in this nonclassified report, the Navy should make every effort to stay abreast of and leverage these developments into its systems.

---

[22]Defense Information Systems Agency. 2007. *Surety, Reach, Speed,* Washington, D.C., March, pp. 26-27. Available at <www.disa.mil/strategy/strategy_book.pdf>. Accessed October 21, 2008.

SUMMARY ASSESSMENT OF INITIATIVES

A major observation obtained by reviewing the work reported on above and referring back to the threat discussion in Chapter 1 is presented in the following finding and recommendation.

MAJOR FINDING: The Department of the Navy has underway a diverse set of IA initiatives that are representative of best commercial IT practices. However:

- No means of integrated assessment exists for determining the impact of implementing the initiatives;
- The implementation of these initiatives will take significant resources and in some cases more than 3 years to implement, leaving a number of naval networks vulnerable to already-known exploitations; and
- Even if all of the existing initiatives are implemented and are successful, these networks are still not assured against the different and more sophisticated attacks that are likely to occur.

MAJOR RECOMMENDATION: Because of the immediate and increasingly sophisticated nature of cyberthreats, the Office of the Assistant Secretary of the Navy for Research, Development and Acquisition (ASN[RDA]), in collaboration with the Office of the Secretary of Defense and the National Security Agency, should conduct a thorough examination of technical opportunities and architectural options and develop a comprehensive plan for reengineering naval networks and computing enclaves to be resilient through cyberattacks by sophisticated adversaries. This plan needs to go beyond commercial best practices, incorporating advanced technology procedures that have been developed by DOD research agencies, mission assurance concepts, and active defense. The plan should also establish operational metrics, baseline these metrics, and set goals for their improvement.

The cyber task force being promoted by the DON CIO, as noted earlier, raises the important point of the need for integrating strategies and activities across cyberdefense, cyberattack, and cyber exploitation. This point was also made in several other discussions that the committee had that are not publicly releasable. The committee does present in Chapter 3 a brief general discussion on the operational merits of integrating cyber offense with cyberdefense. The committee also found that the acquisition and development community often view information assurance in isolation. On the operational side, the committee found that the integration point was well appreciated at the upper echelons, but less so at the lower echelons. These observations led to the following finding and recommendation, which stand as a major overall theme in this report.

FINDING: Information assurance can no longer be treated as an isolated subject, as has traditionally been the case.

RECOMMENDATION: Information assurance should be integrated more broadly with mission assurance to achieve the desired effects—that is, maintaining the availability of networks and the integrity of data and at the same time establishing a broad set of approaches for fighting through successful attacks. The defensive capability provided by information assurance should be supported and augmented by cybersurveillance and cyberattack—just as defense in "traditional" naval warfighting operations is integrated with surveillance and attack.

The remaining chapters of this report elaborate on these findings and recommendations and provide additional recommendations for improving matters.

3

Mission Resilience— Viewing the Threat in Operational Terms

The Department of the Navy (DON) has continued to move toward its network-centric operations vision, depending on commercial information technology (IT) solutions as a principal enabler. The evolving combination of people; weapons; concepts of operations (CONOPS); tactics, techniques, and procedures (TTPs); and advancing information system capabilities continues to enhance naval capabilities across a broad range of missions. The use of integrated commercial off-the-shelf (COTS) IT and interconnected network infrastructures for network-centric command-and-control (C2) systems has helped the department gain many advantages (more informed decision making, improved shared situational awareness, improved information sharing, speed of action, efficiency and synchronization of operations, precision, and cost efficiencies). Future plans extend the use of such COTS IT products into combat weapons systems and bring about an increased convergence between and among C2 capabilities and combat weapons systems.[1] The resulting COTS-based capabilities are anticipated to remain at the heart of DON operations and mission capability.

It has been known for some time that these complex COTS-based capabilities are vulnerable to exploitation and attack.[2] As described elsewhere in this report, it is now apparent that potential adversaries are vigorously working to exploit these vulnerabilities in a variety of ways, including the creation of vulnerabilities

---

[1]Examples include the Aegis cruiser's open architecture, which uses commonly available computer resources, and the DDG-1000 (a planned new class of the Navy's multimission ships), which has a single, commercially based network infrastructure supporting all shipboard functions.

[2]For example, see "The State of Offensive Affairs in the COTS World" at <http://www.fastcompany.com/magazine/127/nexttech-fear-of-a-black-hat.html>. Accessed February 26, 2009.

through the global electronics and software supply chains (for example, a foreign adversary embedding malware into a device before shipping).[3] One can categorize the cyber vulnerabilities of military systems by the type of opportunities that these systems provide to adversaries:

- Espionage or theft of intellectual property, and
- Cyberwarfare (attacks on information capabilities to degrade warfighting capability; such attacks can be in the form of denial of service or manipulation of information and, in the extreme, manipulation or denial of weapons systems).

ADDRESSING NIPRNET AND SIPRNET THREATS

Based on presentations to this committee, most of the current attempted network intrusions that the Department of Defense (DOD) is experiencing are focusing on espionage and intellectual property theft. However, it is widely recognized that adversaries with the capability to exploit military systems for information theft can also apply these capabilities to cyberwarfare.[4] The remainder of this chapter addresses the operational response to cyberwarfare from the perspective of mission assurance.

Naval forces are equipped with a variety of communications and information capabilities that are critical to their warfighting capabilities. (A current general layout for such systems and their computer network defense-in-depth structure is shown in Figure 3.1.)

Among the communications networks available to naval forces is the Non-Classified Internet Protocol Router Network (NIPRnet), an unclassified network that, among other things, provides users with access to the Internet. It is widely recognized that the Internet/NIPRnet connection provides an avenue for adversaries to conduct cyberattacks, including denial-of-service attacks.[5]

The widely reported 2006 cyber penetration that disabled the Naval War College's information network is but one such example.[6] Today, loss or degrada-

---

[3]Defense Science Board. 2007. *Mission Impact of Foreign Influence on DOD Software*, Office of the Under Secretary of Defense for Acquisition, Technology, and Logistics, Washington, D.C., September.

[4]See Jason Sherman, 2008, "DOD Draws Lessons from Cyber Attacks Against Georgia," *Inside Defense,* Washington Defense Publishers, November 10; and John Markoff, 2008, "Before the Gunfire, Cyberattacks," *New York Times,* August 13. Also, Office of the Secretary of Defense, 2008, *Annual Report to Congress: Military Power of the People's Republic of China,* Washington, D.C., pp. 3, 4, and 21, warned that China appears to be aggressively pursuing cyberwarfare capabilities as a key part of its asymmetric "noncontact" warfare strategy. See <http://www.defenselink.mil/pubs/pdfs/China_Military_Report_08.pdf>. Accessed February 26, 2009.

[5]For example, see Erica Naone, 2008, "The Flaw at the Heart of the Internet," *MIT Technology Review,* Vol. 111, No. 6, November/December, pp. 63-67.

[6]"Computer Attack Shuts Down Naval War College Networks," 2006, *Inside Defense,* Washington Defense Publishers, Washington, D.C., November.

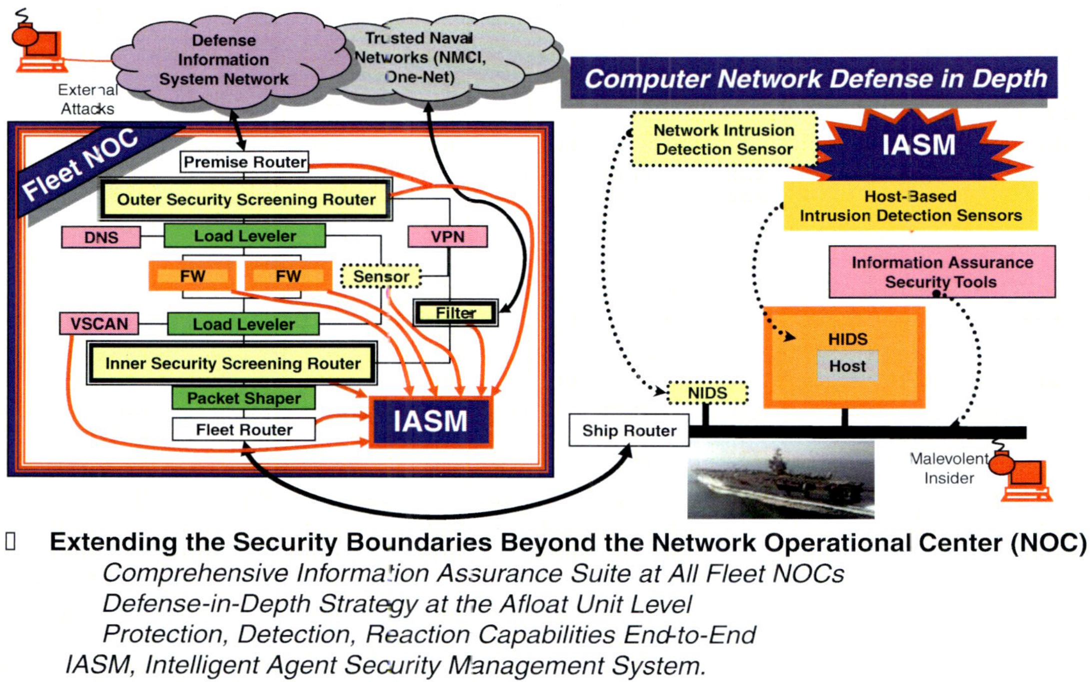

FIGURE 3.1 Naval forces' defense-in-depth computer network defense for shore and afloat infrastructure. SOURCE: Michael Davis, Space and Naval Warfare Systems/Program Executive Office for Command, Control, Communications, Computers and Intelligence (SPAWAR/PEO C4I) Public Presentation, "Information Assurance, What Every Manager Needs to Know," January 10, 2008. Available at <http://www.afcea-sd.org/docs/smallbusiness/IA_Security%20overview1a%20Rev%201_Mike%20Davis.ppt>. Accessed November 10, 2008. NOTE: Acronyms are defined in Appendix A.

tion of the NIPRnet aboard ship and ashore, through a denial-of-service attack, would degrade operations. The primary warfighting areas impacted by NIPRnet loss would be logistics and administrative capabilities.[7] However, closed radio-frequency voice and data communications networks supporting, for example, air wing (aviation) and expeditionary combat capabilities are physically separated from the NIPRnet and would not necessarily be directly impacted by NIPRnet loss. Navy ship crews as well as Marines indicate that they could work around a NIPRnet loss by shifting many NIPRnet users and capabilities onto other available on-ship information networks, such as the Secret Internet Protocol Router Network (SIPRnet), the Joint Worldwide Intelligence Communications System (JWICS), tactical data links, and secure single-channel radio and secure voice systems. However, such shifting would take time and prior coordination, would use channel capacity that may have been designated for other uses, and would likely be effective for a limited time period. If these alternatives are to be used, standardized CONOPS and procedures must be developed across all naval forces and supporting organizations that recognize and practice communications workarounds and autonomous operations with organic sensors on a regular basis. In addition to allowing those involved to "practice as one will fight," these procedures would serve to better inform the operational forces of the true impact of denial-of-service attacks and, through practice, would likely result in better backup procedures.[8]

Successful attacks on SIPRnet/JWICS could be much more debilitating than loss of the NIPRnet would be. Today's networked force relies on the SIPRnet and JWICS for a host of primary warfighting functions, including secure command and control, shared situational awareness, synchronization of joint efforts, access to imagery and other intelligence, mission planning and execution, precision targeting, fires, and battle damage assessment.

An example that raises concern is based on committee briefings and public reports which suggest that the DOD's NIPRnet used by naval forces has been penetrated. A recent report to Congress from the U.S.-China Economic and Security Review Commission states: "*China can access the NIPRnet and views it as a significant Achilles' heel and as an important target of its asymmetric capability*

[7]While the impact of the loss of the NIPRnet over time should not be minimized, most immediate real-time warfighting capabilities reside on the Secret Internet Protocol Router Network (SIPRnet) and Joint Worldwide Intelligence Communications System (JWICS). One of the potential consequences of a denial-of-service attack on the NIPRnet is that much of the traffic that ordinarily rides this network might revert to the SIPRnet or JWICS for those users with access to these networks. This could result in traffic-flow congestion on those networks, forcing them to operate at a reduced capability unless some form of network traffic control was imposed.

[8]Potential new backup procedures should also explore approaches for exploiting those parts of the naval forces structure that can offer potential resilience and restoration benefits, be they the submarine force and its covert capabilities to carry out special functions or the nuclear-powered vessels with the ability to operate autonomously for long periods of time.

[emphasis added]."[9] It also appears that classified networks such as the SIPRnet face many of the same risks as those confronting the NIPRnet.[10]

On the basis of presentations that it received, the committee holds the view that discussions on information assurance (IA) policy across many sectors of the DON are currently centered on how to manage and protect information on networks without reference to the actual use of the information—that is, IA protection policies are not sufficiently being related to the criticality of operations being supported. In the view of the committee, it is important to understand the inventory of mission-critical functions residing on the NIPRnet, SIPRnet, JWICS networks, and the Internet, as well as to assess and understand the consequences of the reduced warfighting capabilities that would result should these networks and systems become degraded. Obvious questions are raised, such as, What is the impact on logistics and other warfighting capabilities should there be a major event that denied access to the NIPRnet and Internet? For example, according to current operational procedures, support contractors, suppliers, and logistics information must be able to directly access the Internet to do their jobs. One must also consider the operational significance if, for example, logistics support was diverted to an unintended location through malware's tampering with network-based information.

In addition to the risk of Internet Protocol (IP) network attacks, today's TDLs (such as Link 16), and secure single-channel radio links, including secure satellite communications, are also potentially at risk. However, these networks are believed to be more secure because of the closed nature of their architecture and—except in the case of a potential kinetic attack—would likely continue to operate independent of cyber events associated with IP networks (such as the NIPRnet). Of particular concern through the next decade is the need for both more broadly available and more-protected satellite communications capabilities to support users without terrestrial connections such as ships afloat.[11]

---

[9]*2008 Report to Congress of the U.S.-China Economic and Security Review Commission,* 110th Congress, 2d Session, November, p. 166. Available at <http://www.uscc.gov/annual_report/2008/annual_report_full_08.pdf>. Accessed February 26, 2009.

[10]For example, the *Los Angeles Times* reports that at least one highly classified network was compromised in a recent severe malware attack at the DOD. See Julian Barnes, 2008, "Cyber-Attack on Defense Department Computers Raises Concerns," *Los Angeles Times,* November 28.

[11]In a related effort, the Transformational Satellite Communications (TSAT) System has been proposed by the DOD; if delivered as currently specified, it would provide military services with high-data-rate military satellite communications and Internet-like services. Touted by the DOD as the spaceborne element of the Global Information Grid (GIG), the TSAT system of satellites is intended to extend the GIG to users without terrestrial connections, such as naval afloat forces, and has been projected by the DOD to vastly improve satellite communications for the warfighter. However, the TSAT program has been not been fully funded by Congress, and the date of its availability is uncertain. (Andy Pasztor, 2008, "Pentagon Delays Program to Build New Satellite System," *Wall Street Journal,* October 21, p. 7, reports that TSAT has been indefinitely delayed.) Meanwhile, the Navy purchases a significant portion of its bandwidth from commercial satellites.

An important example of the dependence on widely available secure communications is the emphasis by the Marine Corps on the use of commander's intent and mission-type orders. The loss of communications would have a detrimental impact on the operating capability and effectiveness of the Marine Corps, particularly when it is working with joint and coalition forces. In the committee's view, for mission resilience the Marine Corps needs to consider establishing multiple diverse reach-back facilities, where the operating forces can access "protected" enclaves of key protected data such as intelligence and logistics information that are critical to the mission. The Marine Corps also needs to conduct an end-to-end review of the original sources of its information to determine the vulnerability of those sources to denial of service or misinformation insertion. In another example, the committee's discussion with representatives of the Pacific Fleet indicated a relatively recent move to place a strong focus on this mission-resilience topic and the assurance of continuous "last mile" connectivity. The Pacific Fleet initiative is a good exemplar for the DON at large.[12]

As discussed in Chapter 2, threat data, coupled with the importance of information to network-centric warfare, have caused the DOD and the DON to consider new IA management arrangements and to set in motion new initiatives related to IA. Threats and attempted intrusions across all DOD networks are documented to be growing rapidly in both number and level of sophistication over recent years.[13] However, from an operational point of view, because the IT networks are considered to be central and critical to the warfighting mission of naval forces, the committee finds the pace of implementing solutions to the growing threat to be inadequate.

Confounding the ability to assess the vulnerabilities and consequences of attacks on naval systems is the myriad of hardware and software configurations that are in use, especially in the cases of legacy systems that may not have the latest security updates or that may lack the proper C2 security structures. The Navy's Cyber Asset Reduction and Security (CARS) initiative[14] will assist in

[12]Examples briefed to the committee by the Pacific Fleet for robust network capability include the application of split IP. In this approach, end-to-end, two-way communication is accomplished through the use of a narrowband highly protected uplink such as Military Strategic and Tactical Relay (MILSTAR) and a robust wideband downlink such as the Global Broadcast System to complete IP transactions.

[13]For example, the committee has been briefed on data showing that across the Navy sensor grid in 2007 there were hundreds of thousands of alarms characterized as high-level alarms, which, after analysis, generated hundreds of reportable incidents or events. Approximately 10 percent of these reportable events were found to have been caused by actions generally attributed to sophisticated adversarial activities. (CAPT Roy Petty, USN, Commanding Officer, Navy Cyber Defense Operations Command, "Overview of Navy Cyber Defense Operations Command," presentation to the committee, April 28, 2008, Norfolk, Va.).

[14]Directed by the Chief of Naval Operations (CNO), CARS is a Navy-wide mission under the operational direction of the Naval Network Warfare Command, assigned to reduce the Navy's total ashore IT assets that are classified as secret or at a lower level by at least 51 percent by September

improving this inconsistent posture by reducing the number of legacy networks across naval forces, providing an inventory of their use, and improving total system security through reducing the potential entry points for external threats. In addition, from a material development perspective, new systems are being developed throughout the DON that are crucially dependent on software for their operation. Even if these systems are not intended to operate on the Global Information Grid (GIG), many may be connected to it for support functions (logistics, maintenance, or training), creating a potentially significant source of IA vulnerabilities.

Because of the immediate nature of the threat to critical information shared on the NIPRnet and legacy networks, the committee recommends that the following mitigating actions be initiated immediately.

MAJOR FINDING: Naval operations are highly dependent on information derived through all networks, including the Non-Classified Internet Protocol Router Network (NIPRnet) and legacy networks. The committee has seen evidence to suggest that the NIPRnet and legacy networks are highly vulnerable, and yet mission-critical functions such as managing logistics are being conducted on these shared networks.

MAJOR RECOMMENDATION: To help address and reduce current perceived network risks related to the NIPRnet and legacy networks, the Department of the Navy should carry out the following:

- Undertake a systematic risk analysis to understand the mission impacts that could be created by information assurance failures. This analysis should be based on an understanding derived through appropriate doctrinal, operational, procedural, and technical analyses—of the information and applications that reside on the networks and how they contribute to mission success.
- Evaluate the implementation of controls that balance operational security risks in posting information on the NIPRnet with the need for information sharing.

2011 and to improve IT security, interoperability, and return on investment. Additionally, by December 2008, it is planned that CARS will deliver full insight into the Navy's total IT asset inventory and the costs associated with delivering and maintaining business and warfighting IT systems and networks. Charles Kiriakou, Head, Cyber Asset Reduction and Security Solution and Security Division, NETWARCOM, "Operations Cyber Asset Reduction and Security, Excepted Network IA/CND Suite Strategy," presentation to the committee, April 28, 2008, Norfolk, Va. (A January 2009 update of CARS reported that of the 1,200 individual Navy networks present when CARS was initiated, only 350 remain to be terminated. Also, during 2008, the CNO accelerated the mission completion time line from September 2011 to September 2010 and raised the bar for total network reduction from 51 percent to 90 percent. SOURCE: Naval Network Warfare Command. 2008/2009. *InfoDomain,* a publication of the Naval Network Warfare Command, Winter, p. 26.)

• Begin to design, architect, and implement the Department of the Navy's networks and systems with an objective of better separating the functions of mission-critical command-and-control systems, logistics, supply, and welfare and morale systems in such a way that an IA compromise in one of these functional areas does not create an IA compromise in others.

• Begin to develop IA operational doctrine that includes being able to conduct mission-critical operations with reduced information capabilities, minimize the time for restoration (reestablishing confidence in capabilities and data), and conduct training exercises for fighting through information attacks, including backup plans for the last mile of connectivity.

LAYING OUT A LONG-TERM OPERATIONAL APPROACH

Operational Response

It is generally recognized that a goal of developing an information assurance capability that would completely eliminate all risk of service disruption and tampering is unrealistic and infeasible. As a result, there is need for a risk-based approach[15] that provides the basis for the DON to develop an integrated cyberattack, exploit, and defend strategy, coupled with a campaign of operational misinformation directed at potential adversaries. However, in addition to adopting a risk-based process for addressing specific IA issues, a well-defined strategy for addressing ongoing network-centric operations is also needed. Taking into account known and projected threat trends, the following operational areas will need ongoing IA-related attention and resources for assured naval network-centric success:

1. *Cyberdefense Concept of Operations.* Naval forces tactics, techniques and procedures for fighting through a cyberattack need to be updated. Such TTPs form the basis for training and exercising against the increased likelihood of such events.

2. *Threat-Based Intelligence Analysis.* There is a need for dedicated, all-source intelligence analysis directed at achieving a better understanding of U.S. adversaries and the threats that they pose—including the intent and capability to develop exploits and the ability to conduct large-scale and sophisticated cyberattacks. A set of directed collection needs must be developed to address important unknowns regarding potential adversaries' intent, and corresponding cyberwar plans must be developed. Results must be coupled with naval mission risk analyses to aid in designing improved mission strategy and tactics, to reduce IA risks, and to maximize the ability to fight through the IA threats. These results would not only be used to stimulate operational responses, but would also stimulate research into the means of achieving the needed collection.

---

[15]Such an approach is described in Chapter 5.

3. *Mission Planning and Analysis.* An effort is needed to model the information dependencies for various naval mission(s), both those conducted solely by naval forces and those conducted as a component part of a joint task force or a coalition force. Models should support the evaluation of degradation in warfighting capabilities owing to current and projected or likely future enemy cyberattack vectors.[16] Mission planning and analysis include (1) the development of integrated cyberattack scenarios; (2) models for exploitative and defensive responses, as well as service restoration strategies and tactics;[17] (3) models related to the use of deception as a cybersecurity strategy;[18] and (4) the use of built-in diversity and fallback strategies and tactics that could permit operation in the face of debilitating cyber and physical activity. Based on these analyses, mission plans and system information assurance requirements need to be developed and prioritized by their impact on the risk of naval forces failing to accomplish their mission objectives.

4. *Minimum and Essential Backup Systems.* Where necessary—and as defined by potential mission risks—naval forces need to be prepared to revert to a minimum essential capability that is as immune to information denial, exploitation, or manipulation as is reasonably possible (analogous to the Minimum Essential Emergency Communications Network used for command and control of nuclear forces).[19] This most basic capability could be as simple as a secure voice-based order wire and/or command wire, independent of normal Internet Protocol networks, augmented with a simple situational display capability. Should new backup systems be developed, consideration should be given to the development of products that are different from the naval standard selections (e.g., different operating systems, different database systems, and so on) to provide diversity that reduces the likelihood of common attack modes.

5. *Resilient Systems.* Naval weapons and information systems, mission strategies, and tactics need to be designed to be more resilient and effective in the face of known and projected IA threats. The Pacific Fleet initiative discussed previ-

---

[16]A discussion of potential kinetic capabilities for disrupting satellite communications can be found in Shirley Kan, 2007, *China's Anti-Satellite Weapon Test,* Congressional Research Service, Washington, D.C., April 23.

[17]Additional discussion on the merits of integrating offensive and defensive cyber operations is included in the final section of this chapter.

[18]A discussion of military deception as a core capability of information operations is provided in DOD Joint Publication 3-13.4, *Military Deception,* July 2006. Available at <http://www.dtic.mil/doctrine/jel/new_pubs/jp3_13_4.pdf>. Accessed February 23, 2009.

[19]The Minimum Essential Emergency Communications Network is designed to provide secure, high-fidelity, jam-resistant, and survivable communications links between the National Command Authorities and the Strategic Nuclear Forces throughout all phases of strategic conflict. Supporting efforts assure an informed decision-making linkage between the President, the Secretary of Defense, and the Commanders of the Unified and Specified Commands. (See Defense-Wide/07 Appropriation/Budget Activity, 2005, Exhibit R-2, RDT&E Budget Item Justification, R-1 Line Item No. 167, February, p. 1. Available at <http://www.dtic.mil/descriptivesum/Y2006/DISA/0303131K.pdf>. Accessed February 20, 2009.)

ously provides a start for a communications system oriented toward addressing this need. This focus on communications systems needs to be expanded to include building resilience for network systems, and it also needs to be expanded beyond the Pacific Fleet.

6. *Training and Exercise.* There is a need for development of significantly enhanced training materials and exercises aimed at improving the proficiency of the DON in utilizing available resources to meet mission objectives in the face of current and projected IA threats. These training materials and exercises should focus on attack recognition and recovery, depending on alternative means of providing command, control, intelligence, and logistics to provide the needed resiliency to successful attacks. They should also include a regular schedule of realistic red- and blue-team exercises based on intelligence estimates of adversary doctrine and CONOPS.

7. *Integrated Wargaming.* Through the years, the Navy and Marine Corps have been leaders in conducting war games that simulate future scenarios and threats. These war games serve to educate and inform current and future leaders on evolving threats, validate naval doctrine and concepts, introduce new and controversial thought, and help formulate budget decisions. The committee believes that expanding the scope of these types of games to include heavy emphasis on cyber operations and mission assurance, using a broad range of cyber experts to formulate the exercises, would serve to better position Navy and Marine Corps leaders to make better operational and IA investment decisions in the future.

8. *Increased Use of More Secure Networks.* There is a need to move critical functions and sensitive information to more restrictive, better-protected communications channels, such as the SIPRnet, where possible. Multiple independent sources for key information elements (to hedge against malicious data manipulation) need to be employed wherever practical. Movement of more information to the SIPRnet may also require movement of the software systems that manage the information. As a general IA practice, only inspected, pristine instances of software packages should be installed on the SIPRnet, and systems currently operating on the NIPRnet should be regarded as infected.

9. *Addition of More Diversity into the Naval Information System Architecture.* Over time, the consolidation of what once were physically separated network nodes and facilities (e.g., satellite terminals, technical control facilities, and so on) has taken place, so as to achieve more efficient and economical operations. This consolidation has often been achieved at the expense of operational diversity. Consequently, there has been the unintended creation of network-wide single modes of failure that could have major direct impacts on operations. The committee believes that an end-to-end review of the current and planned network architecture (to include the IA-related weapons platforms and centralized information nodes) is in order. This review should include a risk assessment of the total command, control, communications, computers, and intelligence (C4I) infrastructure, supporting a

prioritization of investments that would add diversity to the overall naval system architecture.

10. *Reduction of Risks Related to Insider Threats.* Cybersecurity threats from insiders can pose an IA challenge for even the most secure network system. In addition to concerns associated with the potentially harmful accidental actions of insiders, lessons from past insider malicious actions in naval systems are also instructive.[20] The Navy and Marine Corps need to deploy insider monitoring capabilities to detect malicious (or poorly trained) insiders operating within their privilege—or suborned computers operating with legitimate user privilege—but conducting activities outside their normal responsibilities or outside their established and approved patterns of behavior. It should be possible to leverage ongoing activities in counterintelligence and law enforcement to further develop tools for effective monitoring. Insider monitoring can also be extended to correlate physical usage issues (such as accessing computing enclave areas at odd hours) with computer usage.[21]

11. *Improvement of Attribution Capability.* Better capabilities are needed to enable confident attribution of attacks to sources, thereby permitting the initiation of stronger responses, when appropriate, from both a political and a military perspective. In addition, better attribution could potentially serve to facilitate legal recourse.

12. *Updating of Cyberwar Doctrine.* Both the Navy and the Marine Corps need to review their warfighting doctrine and concepts to ensure that the actions, effects, and reactions to computer network attacks, including computer network defense and computer network exploitation, are adequately addressed in the appropriate documents. Policies and lines of authority in these areas must be unambiguous.

---

[20]For example, see Laura J. Heath (Georgia Institute of Technology), 2005, "An Analysis of the Systematic Security Weakness of the U.S. Navy Fleet Broadcast System, 1967-1974, as Exploited by CWO John Walker," Master of Military Art and Science Thesis in Military History, Army Command and General Staff College, Fort Leavenworth, Kans., September 14. Available at <www.stormingmedia.us/03/0396/A039634.html>. Accessed February 24, 2009.

[21]Recent reports describing strategies for insider risk mitigation include *Insider Threat Study: Illicit Cyber Activity in the Government Sector* (Eileen Kowalski, Tara Conway, Susan Keverline, and Megan Williams of the National Threat Assessment Center, U.S. Secret Service, Washington, D.C., and Dawn Cappelli, Bradford Wilkie, and Andrew Moore of the CERT® Program, Software Engineering Institute, Carnegie Mellon University, Pittsburgh, Pa., January 2008); and *Comparing Insider IT Sabotage and Espionage: A Model-Based Analysis* (Stephen R. Band, Dawn M. Cappelli, Lynn F. Fischer, Andrew P. Moore, Eric D. Shaw, and Randall F. Trzeciak, Technical Report CMU/SEI-2006-TR-026, ESC-TR-2006-091 CERT® Program, Software Engineering Institute, Carnegie Mellon University, Pittsburgh, Pa., December 2006). These studies, along with additional case analysis, statistics, and best practices related to insider threat reduction, are available at <www.cert.org>. Accessed February 26, 2009.

Culture Change

A significant change in organizational outlook regarding the importance of information assurance is required if the recommendations of this report are to be meaningfully addressed. In turn, an organizational culture change in the way that systems are built and operated is a key to achieving important IA risk reduction. The committee recognizes that achieving desired improvements in IA will require substantial time and effort and that it is thus important to get the necessary organizational realignment efforts started as soon as possible. In the view of the committee, the required efforts include addressing the following subjects:

- Raising awareness of the cyberthreat by educating, training, and sensitizing the entire leadership and the general workforce to the importance that senior leadership attaches to information assurance in naval organizations;
- Designating senior officers who are held accountable for protecting the valuable information resources that support individual naval operations; responsibilities should be unambiguous, with commensurate authorities (see Chapter 6 for a discussion of IA organizational authority and recommendations);
- Regularly reviewing and/or revising information policies to ensure that they are clear and commensurate with the current threat, state of technology, and operational importance of the information assets that they govern;
- Providing the educational courses, tools, and skills that senior officers need in order to make informed risk management decisions regarding the tradeoffs between IA and opportunities to improve the efficiency of a network infrastructure. The underlying risk analysis methods must be designed to standard assumptions across platforms, so that judgments of individuals are normalized across the Department of the Navy;
- Extending the view that IA and cyberdefense need to employ the full suite of available tools, including deterrence, deception, resilience, and continuity of critical operations while under attack;[22]
- Raising the bar of entry against any adversary attempting to introduce vulnerabilities throughout the system life cycle and supply chain. Given the growing reliance on foreign commercial IT partnering and outsourcing, this threat has become increasingly likely (see the Chapter 4 sections entitled "Architectural Views for Navy Information Assurance Risk Mitigation" and "Information Assurance Research and Development" for technical strategies for raising the bar of entry).
- Developing and deploying deterrents and deception techniques in order to increase the difficulty of exploitation.

---

[22]Joint Staff (LTG Walter L. Sharp, USA, Director). 2006. Joint Publication 3-13, *Information Operations*, February 13, provides further guidance for military information operations planning and execution in support of joint operations. Available at <http://www.fas.org/irp/doddir/dod/jp3_13.pdf>. Accessed February 23, 2009. The committee also argues in Chapter 6 for organizational changes to help drive the DON's integrated approach.

FINDING: Given the current trends related to increasing information system vulnerability, naval forces face significant and growing risks of being unable to execute assigned missions.

RECOMMENDATION: The Department of the Navy should undertake a systematic effort to understand, assess, and strengthen mission capability in light of threats to communications, networking, and information processing systems. This effort should be threat-based; it should include increased operational training and exercises to improve proficiency in working through degraded information environments, using advanced red teams to represent adversarial actions; and it should emphasize educating, training, and the holding of commanders accountable for the protection of the information and networks over which they have responsibility.

INCREASING LEVELS OF INTEGRATION AND SUPPLY CHAIN RISKS

The Department of the Navy's ongoing movement toward integrating information networks (such as the NIPRnet and SIPRnet) with combat weapons systems increases the risk of cyberattacks' disrupting of weapons systems as well as command-and-control functions. In addition, the Navy's open-architecture approach, which uses commonly available commercial products as the computing infrastructure for weapons programs such as Aegis, increases the vulnerability to supply chain attacks of the type described in Chapter 1.[23]

Commercial electronics hardware and software supply chains are increasingly subject to the possibility that adversaries will intentionally incorporate vulnerabilities into hardware and software somewhere in the life cycle between the original equipment development, manufacture, and shipment and the procurement of replacement parts. The risk is greatly exacerbated by the global sourcing of IT hardware and software development, manufacturing, and fielding that takes place today.[24] Currently, nearly all key components used in commercial IT products are developed abroad, with many developed in China, for example.

Recognizing the supply chain risk, the DON may need to revert for certain critical mission applications to a much more trusted supply chain, which could lead to unattractive cost and availability implications. Solutions to resolve this issue should be the focus of naval policy analysis that reconciles cost and other adverse implications with the corresponding reductions in mission risks. In addition to adopting a risk management approach to the supply chain, some specific

[23]Commonly available commercial computers and infrastructure, planned for use in weapon systems such as the DDG-1000 and the next-generation Aegis may contain malicious functionality, which increases the risk that weapons systems may not perform as expected in combat.

[24]For example, see Brian Grow, Chi-Chu Tschang, Cliff Edwards, and Brian Burnsed, 2008, "Dangerous Fakes," *BusinessWeek*, October 2.

operational mitigation techniques for reducing supply chain risks are suggested below:

- Know the provenance of suppliers,
- Protect purchasing information,
- Hide the buyer's identity,
- Have a diverse set of suppliers,
- Mandate transparency in design and manufacturing for buyer protection,
- Limit access for external maintenance and service providers in order to make this avenue of modification harder,
- Minimize the time required between the decision to purchase an item from a particular supplier and the delivery of the item in order to shorten the adversary's window of opportunity,
- Implement trusted distribution processes, and
- Test components after upgrading to increase the odds that a covert modification will be found.

While the IT software and hardware supply chain issue is a DOD-wide issue, the DON should be aware of the concerns and the mitigating operational actions listed above as it develops and implements new mission-critical systems. A recent Defense Science Board report discusses many of these supply chain concerns and outlines potential courses of action for the DOD enterprise-wide organization.[25]

FINDING: In light of current and evolving IA threats, the trends of increasing functional integration and reliance on commercial off-the-shelf IT represent a significant increase in risk to mission operations going forward.

RECOMMENDATION: The management of evolving IA risks requires more attention from the Department of the Navy. For example, the committee believes that it is important to maintain physical separation between the command-and-control information networks (for example, the NIPRnet and the SIPRnet) and combat weapons systems (such as the Aegis, F/A-18, F/A-35, and others). This would reduce the risks of weapons systems being adversely impacted by Internet Protocol network attacks. The committee recommends that the risks associated with the current trend toward highly converged network infrastructure be examined in the context of evolving cyberthreats, including both network-borne and supply chain risks, and that mitigation techniques be developed to address these risks.

---

[25] See Defense Science Board, 2007, *Mission Impact of Foreign Influence on DOD Software,* Office of the Under Secretary of Defense for Acquisition, Technology, and Logistics, Washington, D.C., September, for a more detailed discussion of issues raised by the growing use of COTS products developed offshore and DOD programs currently underway to address the associated assurance issues.

The committee recognizes that in selected cases, direct connections may be appropriate (such as using Link 16 to connect targeting information from the SIPRnet to strike platforms such as Tomahawk or tactical fighters), provided appropriate cyber risk analysis is conducted and the appropriate interface has been established.

THE HUMAN ELEMENT

As the Department of the Navy becomes more network-centric in both its warfare and business processes, the need for increased expertise in cyber and IA technologies and application areas is critical. The network infrastructure has become a major support element for processes within the department, whether warfighting or support functions. In light of the emerging and evolving threat, the department needs to provide the same level of leadership, management, and resourcing to cyber-related issues that it provides to other critical warfighting technology support areas.[26] Accordingly, the committee views the cyber- and IA-related education and training of officers, enlisted personnel, and civilians as a major challenge that needs to be addressed, with the results having a large impact on the degree of information assurance that naval forces can expect. The challenge is heightened by the fact that this education and training must be accomplished within the overall naval education and training program that supports more than 350,000 people.[27]

Education and Training

For the purposes of this report, the committee uses the term "education" to represent formal college and postgraduate education that is principally directed toward the officer community; the term "training" is used to focus on job-specific process learning that is principally, although not exclusively, acquired by enlisted personnel. The committee believes that there is a current and growing need for increased awareness, education, and training in the DON for information assurance. To meet these awareness, education, and training needs, different approaches are required for various personnel at different levels:

- *Improving awareness* to provide broad exposure to the IA subject to high-level officers and civilians who constitute the leadership and operational team;
- *Education* to provide a deeper understanding of the IA subject for officers, select enlisted personnel, and civilians with careers dedicated to the information

[26]For example, the Navy provides dedicated training, management, and resourcing in its Naval Nuclear Propulsion Program.

[27]U.S. Government Accountability Office. 2006 *Information Technology: DOD Needs to Ensure That Navy Marine Corps Intranet Program Is Meeting Goals and Satisfying Customers,* GAO-07-51, Washington, D.C., December, p. 5.

operations community, and for the research, development, and acquisition community; and

- *Training* to provide process-oriented teaching for dedicated officer, enlisted, and civilian personnel supporting computer network defense and system administration, and satisfying the requirements of DOD Instruction 8570 requiring specific levels of information assurance training throughout the DOD.[28]

Identifying and Supporting the Cyber Workforce

Today, the DON's cyber workforce (or "information operations career force" to use current naval terminology) is a mixture of dedicated and "as-assigned" personnel. It is composed of three distinct segments—officer communities, enlisted ratings, and civilian specialists, who possess the preponderance of appropriate skill sets to deliver information operations capabilities.

For example, in the Navy, the officer designators include the following: 1610 (Information Warfare Special Duty Officer); 6440 (Limited Duty Officer, Information Warfare); 7440 (Warrant Officer, Information Warfare Technician Specialty); and 1600 (Information Professional Special Duty Officer). These are communities in which officers spend their entire careers supporting information operations, afloat and ashore. Additionally, those designated 1320 (unrestricted line officers) can support the area of electronic warfare (EW) when assigned to billets that require this skill if they have been trained. Out of 1,460 total billets for 1610 and 1600 officers, the Navy currently fills 1,196.[29]

Enlisted ratings that support the Navy's information operations career forces include cryptologic technicians (CTs) and information systems technicians (ITs). The CTs have further subcategories, chief among which, in the information operations area, is cyptologic technician, networks (CTN), consisting of operators who play a primary role in conducting information operations. An IT's primary role in the information operations area is computer network defense, in which a person may serve in a key role as a system administrator afloat. In the Navy today, there are 7,805 CTs and 787 ITs, and a small number of personnel from selected other ratings, who perform a computer network defense mission.

The officer and enlisted groupings referred to above are the primary component of the Navy's information operations career force. Other personnel groupings may hold notable, specific information operations expertise, but their specific information operations capability is not considered to be their primary area of

[28]Assistant Secretary of Defense for Networks and Information Integration/Department of Defense Chief Information Officer. 2005. Information Assurance Workforce Improvement Program, DOD Instruction 8570, Department of Defense, Washington, D.C., December 19 (updated 8570.01-M, May 15, 2008).

[29]Patrick McLaughlin, Assistant Deputy Chief of Naval Operations for Manpower, Personnel, Training, and Education, N1B, "Overview of U.S. Navy Information Assurance Related Training and Education," presentation to the committee, June 17, 2008, Washington, D.C.

expertise. Personnel in these groupings are assigned information operations-related tasks as a single-tour assignment, or as a collateral duty alongside other areas of warfare expertise.

A variety of factors must be considered in developing and supporting the Navy's current information operations career force of 1,196 officers and approximately 8,600 enlisted personnel, as cited below:

- Current DOD guidance makes no distinction between computer network defense and information assurance. However, the Navy does distinguish between computer network defense and information specialists (CTNs and ITs, respectively) and is currently preparing proposed changes to the DOD to introduce a difference between the two.[30]
- The Navy/Marine Corps Intranet (NMCI) relies on a contracted workforce for the largest outsourced network in the world, which has approximately 650,000 users; however, the Navy does not appear to have good insight into the personnel specialties within the contracted workforce for the NMCI. Updating the workforce strategy for supporting NMCI's anticipated replacement system, the Next Generation Enterprise Network (NGEN), will be an operational necessity, especially if current plans for managing key portions of NGEN with in-house DON civilian and enlisted IT personnel are to be successfully realized.
- The Marine Corps has developed a baseline training and education program, for both their command, control, communications, and computers (C4) enlisted personnel and officers, that needs to be further developed to meet the current and evolving threat. The Corps has an established enlisted occupation field for networking/communication/technical personnel and information assurance personnel, including opportunities for select enlisted personnel to attend programs granting master's degrees in information assurance. These steps have been taken because the environment of widely dispersed forces and distributed operations creates the need to provide network support to the force structure down to the platoon level, and even below in some cases. The increased warfighting emphasis and mission dependencies that are being placed on networked forces, coupled with the time required to develop a more educated and trained workforce and the rapidly changing technology field, lead the committee to believe that it would be appropriate for the Marine Corps to increase its efforts in IT-related training and education with its enlisted personnel, civilians, and officers.
- With the exception of the few officers attending graduate level programs at institutions, such as the Naval Postgraduate School, for most Marine Corps officers their formal education in information technology stops at the company-grade level. While many officers attain additional training at their own initiative,

[30]Patrick McLaughlin, Assistant Deputy Chief of Naval Operations for Manpower, Personnel, Training, and Education, N1B, "Overview of U.S. Navy Information Assurance Related Training and Education," presentation to the committee, June 17, 2008, Washington, D.C.

such as through off-duty education, this in done on an ad hoc basis with no formal structure. The committee believes that there is a need to afford more C4 systems officers the opportunity to attend postgraduate-level education by establishing a formal, continuing-technical-education program for all C4 systems officers beyond the company grade, thereby providing a strong technical core knowledge in information technology and IA requirements.

- There appear to be important gaps in the understanding of the cyberthreat situation among many senior Navy and Marine Corps personnel. To help address this concern, the committee suggests taking full advantage of the information technology program established by the DON for senior personnel, such as the Navy Flag and Senior Executive Service information technology programs, to address cyberdefense and other IA topics.[31] This will help senior officers better understand what information technology can do for them and what the corresponding IA risks are, while also providing a foundation for developing better policies and operational constructs based on the employment of information technology. The above suggestion is made in addition to the committee's recommendation that the Navy and Marine Corps seek more actively to recruit and develop a cadre of future naval leaders with formal degrees in computer science and related information technologies.

Career Paths

Career paths are well laid out for the dedicated officer and enlisted components of the Navy and Marine Corps information operations workforce. The DON's Strategic Studies Group XXVII has recommended a dedicated cyber unrestricted line-officer community. The Studies Group's long-term vision projects a cyber-based warfare community of equal status with the aviation, surface, and subsurface unrestricted line-officer communities.

The committee views cybersystems to be a critical component of a future commander's warfighting capability—comparable to the propulsion, weapons, and logistical systems. Accordingly, commanders must be thoroughly trained and tested in all aspects of the information systems onboard their ships, submarines, aircraft, unit combat operations centers, and carriers, from both a maintenance and an operational perspective. The commander must be able to include integration of cyberwarfare (defensive and offensive) operational strategies with corresponding tactics into their warfighting operations and plans. For the committee, this means that IA considerations should, in the near future, be included in the training and exercising of officers, as well as in consideration of rotational assignments. Furthermore, proficiency in the art of cyberwarfare should be included as one of the prerequisites for career advancement.

---

[31] U.S. Department of Defense. 2006. *Strategic Plan for Joint Officer Management and Professional Military Education,* Washington, D.C., April 3.

Along these lines, the committee was briefed on work underway at Corry Station (Pensacola, Florida) that is aimed at taking a more strategic and aggressive approach toward addressing cyberdefense workforce development.[32] This program provides career pathways, training and education curriculum, and career progression roadmaps for network cyber warriors—from apprentice through master-level skill sets. The program also defines strategic throughput goals across the Services, growing from today's approximately 400 personnel to double that amount over the next 5 years.[33] The Corry Station program is a joint Services effort, including not only Navy and Marine Corps, but also Army, Air Force, and Coast Guard cryptologic and cyberdefense group participation. The Navy leads the joint effort and should be recognized for its vision in this area, as the committee views the Corry Station program to exemplify the type of strategic workforce development planning needed for future cyber operations.[34]

The committee recommends that the Corry Station program be aggressively supported and funded. In the committee's view, the program would be further strengthened by engaging a set of external advisers to conduct a regularly scheduled review of the program curriculum. Such an external review is especially important to help the Corry Station program keep abreast of fast-paced developments in the cybertechnology world.

MAJOR FINDING: The Department of the Navy's workforce, consisting of officers, enlisted personnel, and civilians, has not been required to possess a uniform, prerequisite set of knowledge and IT-related experience. Today's IA-related threats and trends point to a need for the Navy and Marine Corps to address education, training, and career paths as part of the needed response to the growing IA risks and the growing importance of naval cyber operations. The Navy's Corry Station cyber operations training program provides a strong and positive start toward meeting this need.

MAJOR RECOMMENDATION: The Office of the Chief of Naval Operations (CNO) and the Office of the Commandant of the Marine Corps (CMC) should establish a dedicated cyber workforce strategy to include all elements of personnel management (accession, reenlistment, retention, and assignment). Since cyber-related technology continues to evolve rapidly, the cyber workforce program for

---

[32]Although the committee did not directly address the needs or current composition of the workforce for the civilian professionals, a credible naval cyber workforce strategy must also address the future makeup and competency requirements for this segment of the naval workforce.

[33]Richard Matthews, Chief, National Information Assurance Research Laboratory, National Security Agency, "CNO Workforce Development Projections," presentation to the committee, July 16, 2008, Washington, D.C.

[34]The U.S. Air Force has also recently published its proposal for cyber workforce training and education; see Karen Petitt, 2008, "Cyberspace Career Fields and Training Path," U.S. Air Force Public Affairs Memorandum, Scott Air Force Base, Ill., July 2.

naval forces should also include measures to continuously modernize the Navy and Marine Corps training and education curriculum, including the development of formal relationships with universities and external advisers for guiding and supporting naval needs in cyber education and training.[35]

INTEGRATING CYBER OPERATIONS

In testimony before Congress, General Kevin P. Chilton, USAF, indicated that the United States Strategic Command (USSTRATCOM), through the Joint Task Force–Global Network Operations and the Joint Functional Component Command for Network Warfare, is leading the planning and execution of the National Military Strategy for Cyberspace Operations. In this role, USSTRATCOM will coordinate and execute operations to defend the GIG and project power in support of the national interests. General Chilton also testified: "As we continue to define the necessary capabilities to operate, defend, exploit, and attack in cyberspace, we ask for increased emphasis on DOD cyber capabilities."[36]

Within this context, as the DOD defines policy and capabilities to defend, exploit, and attack in cyberspace as part of the overall cyberspace operations strategy, the DON must continue to ready itself both to receive the greatest naval advantage from such capabilities and to effectively support and be supported by joint functions.[37] In particular, new relationships between emerging cyber offense, exploitation, and defense will be established, requiring underlying concepts for integration, with supporting analysis. For example, integration could include *cyberattack* warriors imparting general knowledge and understanding to the *cyberdefense* warriors, perhaps suggesting specific system vulnerabilities that warrant attention. It may be that cyberattack warriors bring a specific attack goal orientation to the IA plan, whereas cyberdefense brings a possibilities portfolio orientation to the IA plan. Also, it may be that as these capabilities are defined, the *cyber exploitation* warriors can support intelligence collection and analysis regarding insight into what exploits adversaries may use in the future, and the *cyberdefense* warriors support intelligence efforts by pointing to areas of concern based on naval mission risk analyses. Also, cyber exploitation and cyberattack warriors may be able to inform exercises that include emulation of enemy CONOPS and TTPs. As these capabilities are defined, the DON needs to assess the following: the pros and cons of various levels of defend, exploit, and attack operations integration; the mechanisms and procedures that would be most

[35]In developing its cyber workforce strategy, the Navy should consider the personnel practices of the Navy Nuclear Propulsion Program as described in Chapter 6 of this report.

[36]Gen K.P. Chilton, USAF, Commander, USSTRATCOM, public testimony before the Strategic Forces Subcommittee of the House Armed Services Committee, February 27, 2008.

[37]The committee believes that the Navy is well positioned to lead the way on integrated cyber operations with the Naval Network Warfare Command and its subordinate commands, the Navy Cyber Defense Operations Command and the Navy Information Operations Command.

effective; and the appropriate balance of investment among these activities. These assessments should serve to guide the relationships that support the broader DOD activities as well.[38]

Another aspect of integration relates to the multi-Service sensor information and communications network dependencies that specific weapon systems rely on. The naval forces require the support of non-naval information systems assets and must supply comparable information for other Services to use. Satisfaction of this need demands that the configuration of individual naval systems, including support systems from other Services, be accurately known and that sensor information on a system-by-system basis be made available to the Navy and Marine Corps, so that both technical and operational reconfiguration can be dealt with in a timely manner. Similarly the corresponding naval information should be made available to support joint commanders and the other Services as their systems require it.

The committee believes that the DON can contribute certain assets and capabilities to a more strategically integrated cyber operations effort that can add significant value to its own IA operations as well as to the broader DOD joint effort.

As the needs for new integrated cyber-related operational capabilities have expanded, the Department of the Navy's initiatives to create and expand the Naval Network Warfare Command have provided a solid foundation for evolving toward a more integrated approach to IA involving defense, offense, and intelligence. However, the committee sees an important opportunity to build on the existing foundation through the development of new concepts and plans that gain additional advantages through greater integration.

MAJOR FINDING: The four cyberspace IA-related domains of protecting, exploiting, attacking, and intelligence do not appear to be closely integrated in the Navy. In particular, the Department of the Navy does not appear to be aggressively considering and assessing alternatives to gain greater IA advantages through such integration.

MAJOR RECOMMENDATION: The Office of the CNO and the Office of the CMC should consider approaches for reducing the separation and enhancing the integration across emerging offense, defense, and intelligence organizations related to IA.

---

[38]Chapter 6 presents a more detailed discussion of naval forces cyber relationships and interdependencies with DOD joint operations.

4

A Suggested Technical Response to Cyberthreats and Information Assurance Needs

This chapter addresses technical issues associated with information assurance (IA) and suggests technical responses to help address exposure of naval forces to future cyberthreats and unsecure system structures. The committee's view is that the Department of the Navy (DON) needs to enhance its present network architecture plans and IA principles significantly to mitigate the present and future threats to its critical operational information technology (IT) systems, which were discussed in Chapter 1. Dependency on elements of the Global Information Grid (GIG) for certain naval missions, in particular, requires that significant IA-related accommodations be made as part of the naval enterprise architecture (EA). In general, it is understood that commercial off-the-shelf (COTS) systems and communication services, while of tremendous benefit to the Department of Defense (DOD) in terms of developmental speed and economic efficiency, are not typically designed with the levels of assurance expected of critical military applications. Intermediate stages of the GIG IA architecture are likely achievable in reasonable time frames and at reasonable cost, but will require substantial improvements to supplement current IA technologies. Recognizing the expected long lifetime of the GIG architecture and its derivatives, the committee's view is that the DON needs to plan actively for the insertion of emerging technologies as part of its architectural plan.

This chapter also offers recommendations focused on naval research and development (R&D) investments and on the rapid acquisition of new, responsive IA capabilities when they are necessary to counter threats.

ARCHITECTURAL VIEWS FOR NAVY INFORMATION ASSURANCE RISK MITIGATION

Enterprise architecture refers to the practice of applying a comprehensive and rigorous methodology for describing the structure and behavior of an organization's processes, information systems, personnel, and organizational subunits so that they align with the organization's core goals and strategic direction. Although often associated with information technology systems, EA is the highest level of an architecture that relates more broadly to the practice of mission optimization by addressing the business mission architecture, performance management, and process architecture. As applied to the Navy and Marine Corps, the enterprise architecture provides the discipline for managing change and complexity within the naval enterprise, especially with constrained budgets. Without an accepted and broadly leveraged enterprise architecture, agile actions in one part of an enterprise could inadvertently be detrimental to another part.

For the naval forces, EA must be viewed broadly, extending well beyond the traditional boundaries often associated with IT architectures. For the purposes of information security, IA considerations are a critical aspect of the EA. Taking into account key mission areas, IA considerations must include the following:

- How to engage in and respond to information attacks;
- How to deal with insider threat issues;
- What the impact and response mechanism should be for communications system jamming and data-link loss;
- What the placement and use of both physical and cyber intrusion sensors should be for optimum protection;
- How to best detect and mitigate information theft and tampering;
- How to allocate user access to information systems dynamically, to account for rapidly changing circumstances; and
- How to monitor IA performance as a necessary aspect of continuous system improvement as threats evolve.

Today, systems are being integrated without full consideration of the impacts on information assurance, which can lead to unanticipated vulnerabilities[1] and overarching system weaknesses. To counter such vulnerabilities, the EA should be designed according to a set of principles that address the essential characteristics of a system for assuring information. The process for architecture development must be iterative and adaptive to ensure that naval systems will remain robust in the face of emerging threats and evolving technologies. The remainder of this section discusses a set of principles to guide the incorporation of IA throughout a

[1]In some cases vulnerabilities are known but are considered acceptably low risks.

capability life cycle and describes a set of technologies that the naval forces could consider to enhance their assurance.

The Influence of the Global Information Grid Architecture on Current Developments

The missions of all Department of Defense Services are becoming more complex and dependent on information sources distributed widely around the globe. Consequently, the DOD has developed a vision for the GIG. Current plans are to build out a DOD-wide architecture serving the needs of all Services and coalition partners for its information, warfighting, and business needs. It is this broad vision for the GIG to encompass all aspects of naval operations that causes concern to this committee. The GIG's guiding principles are to provide information services anywhere and anytime in an architecture that is extensible, affordable, and assurable, providing the full range of services necessary to support the (joint) missions of each of the Services.[2] Thus, one key issue of concern for this committee is the integrity and the trust of the information networks and the computational systems that constitute the GIG.

Based on a core set of attributes that the DOD and the DON have advocated, a set of principles can be enumerated from GIG architecture descriptions that should guide naval information and computational systems design and development:

- *Design for interoperation across (mobile) platforms.* The basic information architecture should enable interoperability and should be augmented to ensure that information assurance and trust can be achieved. Furthermore, the gateway of services and data should be conducted through the application of high-assurance devices.
- *Encapsulate modules for extensibility.* For a system to operate properly, functions and services should be encapsulated in a manner that supports easy integration into multiple solutions and applications.
- *Isolate application content data from process control information.* Data, applications, and business processes should be separated to ensure isolation of control information from application-related content.

These principles are at the core of object-oriented design and provide extensibility, incremental development, and efficiency through reuse of common services and data. With respect to the defense of the GIG and protection of its critical services and interfaces, the IA architecture aspects of the GIG must also

---

[2]Department of Defense Chief Information Officer. 2007. *Global Information Grid Architectural Vision: Vision for a Net-Centric, Service-Oriented DoD Enterprise,* Version 1.0, Department of Defense, Washington, D.C., June, p. 24. Available at <http://www.defenselink.mil/cio-nii/docs/GIGArchVision.pdf>. Accessed November 17, 2008.

include principles to prevent attack and security breaches, and to detect when security is compromised and to respond. The following principles are suggested by the committee:

- *Segment systems and separate communication pathways into trust levels.* Wherever practical, efforts should be made to segment information systems into tiers corresponding to different levels of needed security. Separation into different tiers must include the establishment of *suitable* controls (technical and procedural) related to accessing higher security tiers from lower security tiers. Technical controls can be software-based and hardware-based, depending on the level of security desired and the specifics of the systems in question. Suitability depends on the risks associated with undesired access occurring and the costs associated with implementing tighter controls (cost can be measured in dollars, system performance degradations, system usability, and other ways).
- *Encrypt channels and enforce (policy-based) access controls at the interfaces.* To help protect the core network and critical services from attack, data and control information on the network should be encrypted and the interfaces to the network should have strict access controls.[3]
- *Audit and archive encapsulated modules for security.* Information services that are part of the core should be implemented as encapsulated software modules, with strong configuration control of their interfaces and with the requirement that they be subjected to a monitoring infrastructure that audits the use or misuse of modules to isolate trouble or IA failure.
- *Advanced IA design principles, IA tools, and IA products should be pervasively deployed at the network level and to the end points and should be continuously refreshed.* To allow rigorous audit and response to the data entering and moving through the GIG, IA tools, such as those being developed and deployed for the Comprehensive National Cyber Initiative, should be installed at the edge and broadly across the critical internal systems and services that constitute the GIG, providing the basis for boundary protection within a layered-defense assurance system. Assurance cannot be guaranteed without also pushing IA tools to the hosts and client machines that use and depend on the GIG.[4]

[3]Encryption policies and technology for sensitive information are well defined in DOD policy documents, such as the National Policy on the Use of the Advanced Encryption Standard to Protect National Security Systems and National Security Information (CNSSP-15), and the classified encryption technologies maintained by the National Security Agency. See <http://www.nsa.gov/ia/programs/suiteb_cryptography/index.shtml>. Accessed February 18, 2009.

[4]One example of such tools is the Host-Based Security System, based on McAfee and other COTS products and being deployed across the GIG by the Defense Information Systems Agency. However, as discussed in Chapter 2, current COTS products do not protect against so called "zero-day" exploits that have not been previously seen. See Secunia, 2008, Internet Security Suite Test, October. Available on the Internet at <http://secunia.com/gfx/Secunia_Exploit-vs-AV_test-Oct-2008.pdf>. Accessed November 14, 2008.

The Dependency of the Global Information Grid on COTS Technologies

The application of the design principles provided above to Navy systems is a critical part of achieving needed information assurance levels. However, without demanding that DON system designs consider IA properties to be of comparable priority with additional system functions, system cost, and system efficiency, the Navy and Marine Corps will be in the position of assuming levels of risk that have not been given sufficient consideration.

Example of the Need for IA Principles to Support Global Information Grid Design Principles

Service-Oriented Architectures

The design principle of extensibility of the enterprise architecture is largely provided by the traditional object-oriented paradigms that have culminated in a service-oriented architecture (SOA) implementation. SOAs enable open, flexible, and adaptable systems and are designed to readily enable interoperability across disparate systems that may be under the control of different ownership domains.[5,6] This architecture also allows for faster integration of mission or business processes, both within the DON as well as across the larger DOD organization. However, SOA can be both an opportunity and a vulnerability for the GIG, as it offers flexibility to evolve capability over a wide class of users, but also offers paths for potentially vicious code to find its way into a warfighting system. Within the DON as well as the larger DOD community, SOA implementations have focused heavily on Web services as the enabling technology. A discussion of SOA approaches planned by the Navy and associated IA design consideration is provided in Appendix E.

SOA as an architectural approach to building distributed systems is not inherently vulnerable; however, implementations of SOA using specific technologies, such as Web services, could require the application of special security protection to mitigate broad threats to the DON and DOD enterprises. For example, in an object-oriented, distributed computing environment such as that supported by SOAs, communicated data can include embedded code that the recipient uses to integrate with another information service(s). The consequence to IA is obvious: an object-oriented framework provides a code-injection platform that not only serves the intended system participants, but also can serve potential attackers who

[5]OASIS [Organization for the Advancement of Structure Information Standards] Open Organization. 2006. Reference Model for Service Oriented Architecture (SOA) 1.0, Billerica, Mass., October 12. Available at <http://docs.oasis-open.org/soa-rm/v1.0/soa-rm.pdf>. Accessed August 22, 2008.

[6]M. Brian Blake. 2007. "Decomposing Composition: Service-Oriented Software Engineers," Special Issue on Realizing Service-Centric Software Systems, *IEEE Software*, Vol. 24, No. 6, pp. 68-77, November/December.

intend to insert their system exploitation code. Without IA design principles that are applied to help mitigate the risks attendant with new commercial technologies, IA can be severely impacted in a variety of undesirable ways.

The capability just described is exemplified in modern computing on the Web and within COTS products generally. Web pages fetched from a server are not passive documents, but rather are complex objects extended with code (such as JavaScript) to render the document's content, including perhaps a rich set of embedded media, in a local client browser. Indeed, modern document formats (Word .doc, and Adobe .pdf) are not passive text files, but are full-fledged computational objects with embedded code that must execute in order to render and display the document at a client machine. When code is injected into any cooperating process, the key IA question is this: *Is that code benign and friendly, or malicious and dangerous?* Irrespective of the manner in which the active data are communicated (whether they are encrypted and protected in transit or not), the data injected into a process may pierce security boundaries by appearing compliant with the interface policy. Consequently, object-oriented computing invites certain IA risks by serving adversaries who have seized the opportunity to inject their malicious code in myriad ways. SOA environments that fully embrace the object-oriented paradigm inherent in Web services thus require special attention from an IA perspective.

Traditional perimeter and host-based security solutions and technologies are limited in their ability to protect Web services, given the dynamic nature of Web services-based SOA environments that often extend beyond the operational and physical boundaries of a single domain or network. This situation is exacerbated because, all too often, many organizations allow Web services traffic to flow, without restrictions, through firewalls, given that they use the same ports and protocols as Web traffic. Although enterprise service buses have been introduced as enterprise-wide containers of Web services, the state of the art in these systems lacks the interoperability policies and protocols required to securely integrate an organization as large as the DON's system of systems. The GIG-influenced architectural design principles envisioned for future naval platforms and systems clearly point toward making use of COTS products within an SOA framework, and possibly cloud computing architectures as well.[7] Hence, the committee believes it to be inevitable that future Navy systems will be subjected to new and more complex IA vulnerabilities presented by the use of SOAs and related COTS products.

[7]Cloud computing is a computing paradigm in which tasks are assigned to a combination of connections, software, and services accessed over a network. This network of servers and connections is collectively known as "the cloud." The concept of cloud computing and acquisition of software as a service for naval forces was not examined in detail by the committee; however, these concepts also carry IA risks and vulnerabilities of which the naval forces should be aware. For a recent article discussing these risks, see Dan Goodin, 2009, "Multi-Site Bug Exposes Cloud Computing's Dark Lining," *The Register*®, March 12. Available at <http://www.theregister.co.uk/2009/03/12/cloud_computing_dark_side/>. Accessed March 23, 2009.

MAJOR FINDING: As part of its implementation of network-centric warfare capabilities, the Department of the Navy is aggressively embracing integrative COTS technologies such as service-oriented architectures in order to take advantage of potential positive benefits, including wider information availability. However, these adaptations also have the potential to introduce new and possibly serious IA risks into naval systems. Unfortunately, existing naval systems do not appear to have been designed with consideration of the collateral IA risks as a foundational system attribute.

MAJOR RECOMMENDATION: In order to provide the appropriate level of information assurance, the Office of the ASN(RDA) should adopt and manage system developments using sets of IA principles that are explicitly specified and required to be incorporated into the naval forces enterprise architecture, including specifically addressing the IA requirements of service-oriented architectures. In addition, these principles need to be embraced throughout the system life cycle and adopted by existing naval systems as they are upgraded.

IA Risks of Current COTS Technologies

Naval mission capabilities are being designed and built today by logically integrating and coupling information systems and capabilities to weapons and sensor systems. Correspondingly, assuring access to and use of this information technology is becoming increasingly critical to avoiding degradation in mission performance. Given the pervasive use of COTS technologies as the basis for these integrations and the questionable state of information assurance that these products provide, defending naval systems is significantly more difficult. It requires protections against opportunities for adversaries to attack using jamming, cyber manipulation, physical attack, malicious code injection, and sabotage anywhere in the network. As described in Chapter 1, new vulnerabilities associated with COTS software and hardware are regularly reported by government agencies established to monitor cyberthreat activities. For example, weekly bulletins listing and ranking newly discovered software and hardware vulnerabilities are provided by the United States Computer Emergency Readiness Team.[8] Additionally, examples of COTS-related vulnerable hardware and software incidents are routinely reported in the public media.[9]

Related to this risk, the long-term vision of the GIG provides for a phased approach to the development of a common communication infrastructure, providing multiple security levels and IA by logically isolating disparate levels (as

[8]A profile of weekly vulnerability bulletins is available at <http://www.us-cert.gov/cas/bulletins/>. Accessed March 19, 2009.

[9]See footnote 27 in Chapter 1 of this report for examples of postings of daily cyberthreats and Internet security news alerts on CyberInsecure.com.

represented in Figure 4.1).[10] In this vision, IA technology improvements are expected, over time, to provide the technical means for safely integrating multiple communications paths with the levels of assured separation required for a rich variety of naval missions. Yet, in the opinion of this committee, for certain critical military applications, such technical capabilities may not be sufficient when relying on today's COTS technology or on that available in the foreseeable future. Given the current state of IA in the commercial marketplace today and the long-term vision of the GIG's communication substrate, as represented in Figure 4.2, the committee believes that—for critically sensitive systems—it is technically prudent to depend on physical isolation of multiple data paths in current Navy communication systems designs, even though one might find various efficiencies in relying on software-based logical isolation.

This situation creates a new set of security design issues that is well represented by the specific cases of the Navy's current design of the DDG-1000 and the Consolidated Afloat Networks and Enterprise Services (CANES) communication subsystems. For example, the logical block diagram of the onboard communication system and its physical layout aboard ship of the DDG-1000 was presented to the committee.[11] The design reveals that the communication channels spanning multiple trust levels, including control of the ship, all flow within the same physical cabling and switching subsystems. The vulnerable COTS networking products cited above serve as the core equipment used in this design. Although the committee did not conduct a detailed review of the IA analyses that were performed to certify and accredit this design, it is concerned by this approach. The design has been certified and accredited, but the approach of relying solely on logical (software-based) isolation of critical networks nonetheless introduces a level of risk that physical separation of specific critical network systems would avoid. The committee believes that the Navy has to decide in its overarching IA policy which levels of risk it is willing to take. The level of risk associated with consolidated critical networks may be appropriate; however, the committee believes that decisions about such levels of risk should not be delegated to system designers to decide, and should include the inputs of a broader set of naval leadership.

Through many decades of evolutionary development, the naval forces have acquired a vast array of information-based systems. In aggregate, these systems address naval forces requirements for communications and networking, data processing, and command and control. With FORCEnet as the context, the challenge now is to get broad user acceptance of the architecture, incorporate various naval

[10]Department of Defense Chief Information Officer. 2007. *Global Information Grid Architectural Vision: Vision for a Net-Centric, Service-Oriented DoD Enterprise*, Version 1.0, Department of Defense, Washington, D.C., June. Available at <http://www.defenselink.mil/cio-nii/docs/GIGArchVision.pdf>. Accessed February 17, 2009.

[11]Myron Liszniansky, DDG-1000 Software Integration Manager, Program Executive Office Ships, "A Cost-Effective Approach to Certification, Test, and Evaluation (CTE)," presentation to the committee, June 18, 2008, Washington, D.C.

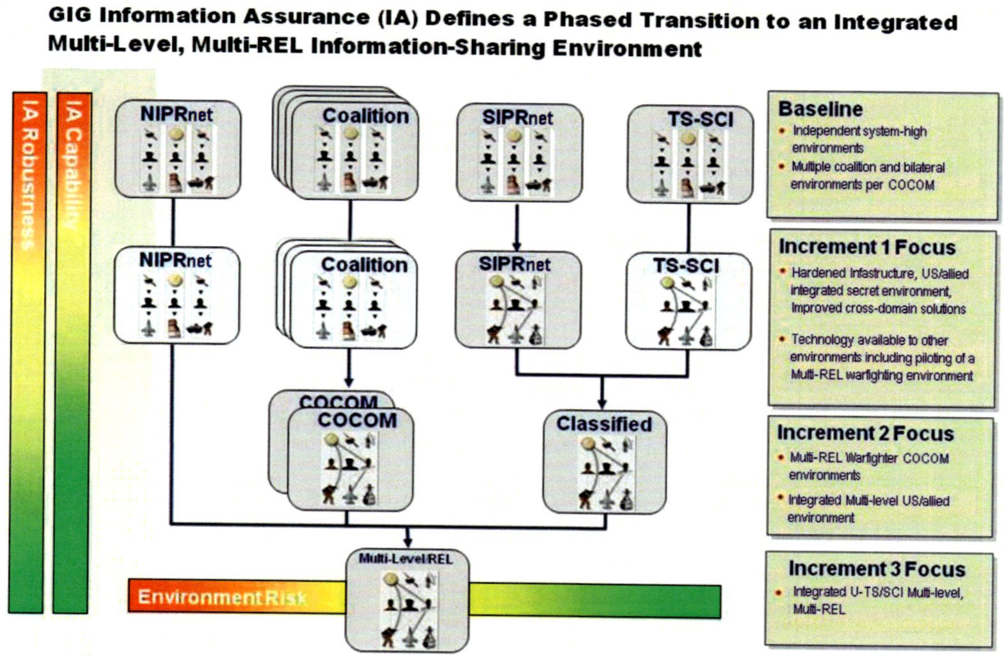

FIGURE 4.1 The long-term vision of the Global Information Grid (GIG) with fully integrated but logically separated multi-tier security channels. NOTE: REL = Rights Expression Language, used to enable the authorized distribution and protection of valuable data. Other acronyms are defined in Appendix A. SOURCE: Adapted from Craig Harber, Enterprise IA Architecture and Systems Engineering Office, Information Assurance Directorate, National Security Agency, 2008, "GIG Information Assurance Architecture, Protecting National Security Enterprises" (viewgraph presentation). Available at <http://www.cisse.info/colloquia/cisse10/briefings10/Harber.ppt>. Accessed October 23, 2008.

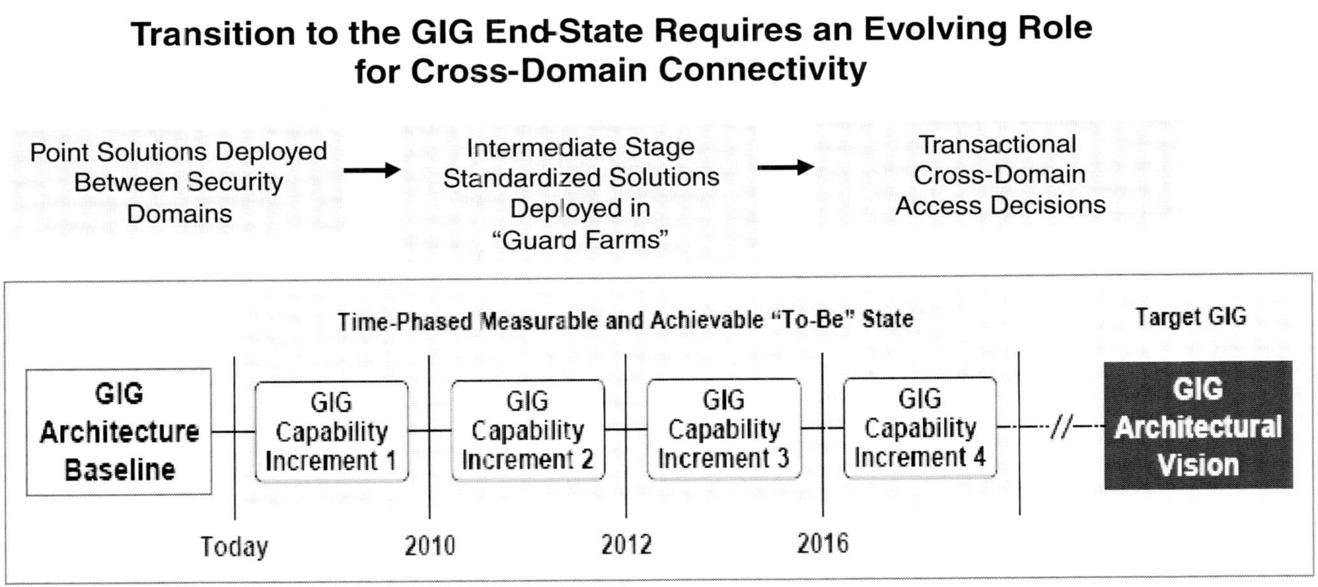

FIGURE 4.2 Phased transition of the Global Information Grid (GIG) architecture includes an intermediate stage that may be achievable in a reasonable time frame (dates shown are notional). SOURCE: Adapted from Department of Defense Chief Information Officer, 2007, *Global Information Grid Architectural Vision: Vision for a Net-Centric, Service-Oriented DoD Enterprise,* Version 1.0, Department of Defense, Washington, D.C., June. Available at <http://www.defenselink.mil/cio-nii/docs/GIGArchVision.pdf>. Accessed November 17, 2008.

assets into the architecture, and ensure that naval capabilities fully embrace the IA principles previously discussed. The architecture should also align with the broader GIG vision, which incorporates many of the core principles.

Although the naval forces have made an excellent start with their "to-be" architecture, the committee's assessment is that successful realization will be challenging. The naval enterprise is as complex and diverse as any commercial or other government infrastructure. Making the situation more difficult are the challenging requirements of naval operational missions relying on global reach, a vast user base, a highly diverse set of platform types, and time sensitivities in an environment in which information attack can cripple operational capabilities.

The designers of the to-be GIG architecture recognized the need for a phased introduction to the long-term vision of a fully enabled, globally accessible SOA as IA matured over time. Figure 4.2 displays this phased transition, showing an intermediate stage that is likely achievable in a reasonable time frame, given substantive and realistic improvements in IA capabilities and a case-by-case reconsideration of COTS dependencies. The ultimate long-term capabilities of the fully developed GIG remain largely out of reach without substantial new breakthroughs in IA.

Although the naval forces can implement security protection on a service-by-service basis, a more effective IA strategy to securing Web-service-based SOAs is to externalize crosscutting security functionality, such as encryption, authentication, auditing, policy enforcement, and so on, into a shared services infrastructure that can be consistently managed, configured, and coordinated by security professionals rather than by individual development teams. An example of a shared security services infrastructure would be the integration of the Navy's CANES Web services implementation with the Net-Centric Enterprise Services (NCES) security services structure developed and deployed by the Defense Information Systems Agency (DISA).[12] It is recognized that there are IA trade-offs associated with externalizing and centralizing security functionality (e.g., concentration of risk or overdependence on a central security organization to be capable of adequately addressing the security and operational needs of a complex distributed system). However, the DON, as part of implementing SOA-based capabilities, should explore the trade-off space to find the most viable solution. In any case, because of GIG-based interdependencies, the Navy and Marine Corps SOAs will need to be developed in very close coordination with the DOD community.

MAJOR FINDING: The Global Information Grid (GIG) architecture promises to provide secure information services that are envisioned to be electronically integrated into weapons systems and other mission-critical control systems. This vision is highly dependent on trustworthy commercial off-the-shelf (COTS)

[12]See DISA's NCES Service Oriented Architecture Foundation services. Available at <http://www.disa.mil/nces/product_lines/soa.html>. Accessed November 14, 2008.

technology components. The Department of the Navy, in keeping with the GIG architecture vision, is increasingly dependent on logical (software-based) information isolation rather than on physical separation for highly integrated, warfighting-critical systems composed largely of COTS components. This strategy is risky from an IA perspective, given the demonstrated vulnerabilities in COTS components.

MAJOR RECOMMENDATION: The Office of the Assistant Secretary of the Navy for Research, Development and Acquisition (ASN[RDA]), in conjunction with other interested Navy and Marine Corps elements, should reexamine its IA architecture and design strategy, with emphasis on establishing the IA worthiness of the current systems under development. Special attention should be given to (1) the IA aspects of isolation and separation inherent in the outcomes in the Navy's Consolidated Afloat Networks and Enterprise Services (CANES) program and (2) the DDG-1000 onboard communication subsystem.

INFORMATION ASSURANCE RESEARCH AND DEVELOPMENT

The State of Naval Forces Information Assurance

Given the committee's concerns that IA should have equal priority in all current and future enterprise architecture designs, and given the state of IA readiness in current COTS systems, the committee believes that the Navy should immediately invest in IA research and development initiatives to remain current and capable of deploying IA solutions protecting the Navy's primary missions.

Information assurance on naval forces networks today is largely managed by implementing best commercial practices. In the committee's review of naval forces information assurance strategies, a number of strengths emerged:

- The naval strategy of applying current best practices in the configuration management of desktop and server systems to ensure that a common configuration of systems is deployed, managed, and regularly patched;
- The naval strategy of deploying both desktop antivirus signatures and gateway signatures for detecting attacks against Navy networks; and
- The Navy's Cyber Asset Reduction and Security (CARS) program to reduce and ultimately eliminate currently unmanaged Navy networks and systems.[13]

These strengths in managing naval forces networks, in the committee's view, are accompanied by significant shortcomings in providing information assurance for naval networks:

---

[13]See footnote 14 in Chapter 3 of this report for information on the CARS program.

• Detection of threats is outward facing. Current sensors are positioned largely to examine Internet-sourced traffic rather than also to examine threats from within naval networks. An inward orientation, in addition to the current outward orientation, is needed to detect threats currently running on and within naval networks, as well as future threats that breach perimeter security.

• Detection of threats for naval network systems is driven primarily by a top-down, centralized command-and-control organizational structure (e.g., Joint Task Force–Global Network Operations → Naval Network Command → Navy Cyber Defense Operations Command → fleets and bases), and by network topologies (e.g., Non-Classified Internet Protocol Router Network [NIPRnet] gateways → Network Operations Centers → bases and fleets → enclaves → local area networks). To relate discovered threats to mission-specific risks, detection processes should also include the integrated monitoring of the information system assets that together perform mission-supporting functions. Currently the top-down orientation for threat detection does not incorporate the service-specific system functional configurations as a factor. To provide a mission-oriented threat detection capability, the current top-down approach needs to be supported by data derived from service monitoring activities that provide detection inputs from a system function and mission perspective.

• Current threat detection is hampered by attack noise. The detection of threats can be enhanced by investment in more effective *prevention* technologies. An improvement in prevention could potentially eliminate the bulk of the conventional attack noise that now clutters detection sensors, allowing detection to be more focused on advanced threats, while also preventing unsophisticated "gateway" threats (e.g., standard botnet-driven attacks[14]) from gaining a foothold in Navy networks.

• Current IA strategies do not sufficiently address either current sophisticated attacks that cannot be handled through use of existing COTS security products, or future projected cyberthreats. Because the cyberthreats that naval networks face include targeted, evasive, sophisticated threats, the DON needs to actively pursue technology to address current and future threats rather than relying entirely on best-practice COTS tools. The research community has provided significant evidence that current signature-based sensors are at present obsolete in detecting current and future projected threats.[15] Pervasive and exclusive use of signatures and patching methodologies leaves current naval systems and platforms exposed and blind to advanced threats.

---

[14]A "botnet" is commonly defined as a network of independent programs, or bots, acting in concert.

[15]For example, see Yingbo Song, Michael E. Locasto, Angelos Stavrou, Angelos D. Keromytis, and Salvatore J. Stolfo, 2007, "On the Infeasibility of Modeling Polymorphic Shellcode for Signature Detection," *Proceedings of the 14th ACM [Association of Computing Machinery] Conference on Computer and Communications Security*, ACM, Alexandria, Va., pp. 541-551.

- Insufficient attention is given to insider threats in the current naval IA strategy. Insider threats can be far more sophisticated and effective than remote attacks. The lack of insider threat detection strategies and tools leaves a major capability gap in protection of naval networks.

Addressing these IA shortfalls will require the DON to invest in an integrated, advanced-research, rapid-deployment approach to IA—an approach that can operate on the same time lines as those of adversaries intent on attacking naval systems. While the initiatives of the DOD and the Comprehensive National Cybersecurity Initiative (CNCI)[16] will be useful to the overall IA posture of naval forces, these initiatives by themselves will not address important naval-specific needs such as those enumerated above.

Comprehensive National Cybersecurity Initiative

In January 2008, it was announced that President George W. Bush had approved a plan and submitted a budget to Congress for the Comprehensive National Cybersecurity Initiative, and that the U.S. intelligence community (IC) would have a major role to perform in the CNCI's execution.[17] The committee is aware that the IC has very substantial capabilities that could be brought to bear in the effort to achieve cybersecurity, but it is also cognizant that the IC has statutory roles and responsibilities different from those of the Navy and the other Services.[18] Further, the participation of the IC in the CNCI has already generated controversy in Congress and the press. Consequently, the Navy should not assume that its cybersecurity needs will be accomplished solely through the CNCI; instead, the

[16]The CNCI was officially established in January 2008, when President George W. Bush signed National Security Presidential Directive 54/Homeland Security Presidential Directive 23: "Cyber Security and Monitoring." The CNCI is a multiagency, multiyear plan that lays out 12 steps to securing the federal government's cybernetworks.

[17]At the time of this writing, many aspects of the proposed CNCI are uncertain; for example, see Senate Committee on Homeland Security and Government Affairs, May 2, 2008, Press Release: "Lieberman and Collins Step Up Scrutiny of Cyber Security Initiative." Available at <http://hsgac.senate.gov/public/index.cfm?Fuseaction=PressReleases>. Accessed January 29, 2009. Also, as an update to related events since the change of administrations in January 2009, it is reported that President Barack Obama has ordered his national security advisers to begin a comprehensive, 60-day review of federal cybersecurity initiatives as a prelude to developing an integrated cybersecurity strategy across federal agencies. (Office of the Press Secretary. 2009. *President Obama Directs the National Security and Homeland Security Advisors to Conduct Immediate Cyber Security Review*, Press Release, The White House, Washington, D.C., February 9.) The status of the Comprehensive National Cybersecurity Initiative and its associated strategic research thrust will likely be heavily influenced by the yet-to-be-released outcome of this review.

[18]Although the DOD and National Security Agency cyber operations are governed primarily under *United States Code*, Title 10, authority (laws governing federal military action), other specific IC-related aspects of cyber operations are governed under *United States Code,* Title 50, authority (laws governing foreign intelligence activities).

DON needs to be prepared to defend its own portions of the GIG as advised in this chapter.

Advanced Research and Development Strategy

For the reasons outlined above, the committee believes that it is necessary for the DON to significantly increase its own science and technology program in IA[19] and to develop relationships with a sufficient number of leading researchers focused on IA. The Navy's own science and technology program can be grown quickly by leveraging ongoing advanced R&D programs at the Defense Advanced Research Projects Agency (DARPA), Air Force Office of Scientific Research (AFOSR), Army Research Office (ARO), Intelligence Advanced Research Projects Activity (IARPA), National Security Agency (NSA), National Science Foundation (NSF), Department of Homeland Security (DHS), other federal R&D agencies, University Affiliated Research Centers, Federally Funded Research and Development Centers, and industry investments as appropriate. Leveraging advanced R&D from others could bridge the technical capability gap, enabling the DON to potentially leap ahead of the current cyberthreat and better position naval forces against a clear and present cybersecurity danger that threatens the ability of the naval forces to execute their missions.

Given the trends in military information technology and networks, the current and growing sophistication of potential adversaries in cyberspace, the current posture of DON information assurance, and the capability gaps in defending against the cyberthreat, the committee recommends a double-pronged naval IA research strategy: (1) invest in cybersecurity research (a) to address naval-specific capability gaps and develop a robust research program that is relevant to the Navy and that may not be currently addressed elsewhere, and (b) become an active participant in the IA research community to develop the knowledge and relationships needed to rapidly transition technology; and (2) establish a rapid technology testing and evaluation laboratory and a technology insertion program to leverage and accelerate ongoing research in cybersecurity into Navy networks.[20] The committee recommends below that the Office of Naval Research (ONR) play a leading role in IA research and development to establish the knowledge base and intellectual property of the Navy and Marine Corps for insertion of IA into naval systems.

---

[19]In conformance with the Secretary of the Navy Instruction (SECNAVINST) 5239.3A, from the Secretary of the Navy to All Ships and Stations re: *Department of the Navy Information Assurance Policy,* Department of Defense, Washington, D.C., December 20, 2004, p. 15, item 3.

[20]In 1999, the Department of the Navy adopted a new process for concentrating its scientific and technological resources to prepare for the Next Navy. The Future Naval Capabilities (FNC) process shifted the science and technology investment focus from individual technology goals to the most highly desired future capabilities for naval forces. Additional FNC information is found on the Internet at <www.nrl.navy.mil>. Accessed November 12, 2008.

Suggested Elements of an Advanced Research Program

The committee reviewed a number of previous R&D surveys in IA, including a 2006 federal interagency report,[21] an Air Force-commissioned study,[22] a Defense Science Board report,[23] and previous cybersecurity R&D reports from the National Research Council.[24] The committee's view is largely in agreement with many of the findings and conclusions in those studies. By way of summary, the severity of current and future threats has been identified, and research activities have responded to these threats by studying new concepts in system design and security functions. Broadly, active research is ongoing in the following areas:

- Secure network functions (routing, addressing);
- Computing systems that are of high integrity and trusted;
- Survivability and recovery of networks after large-scale attack, including rapidly reestablishing trust;
- Secure compositions of complex systems composed of insecure elements;
- Trustworthy platforms and secure application designs;
- New authentication and access control systems to protect the privacy of data and restrict access to critical systems; and
- New models and metrics for security.

Nonetheless, based on the committee's review of these studies and on briefings from Navy personnel responsible for the Navy's R&D initiatives,[25] a number of capability gaps in network, system, host, user, and privileged-user security became apparent. There are existing federal government research investments in these areas, but the Navy needs to be a stronger participant in the IA research community to help ensure that these efforts are directed at Navy-specific needs and to be able to understand and leverage the research results. The briefing on naval research efforts from the Office of Naval Research to the committee revealed that ONR has limited resources and few programs in IA that address important aspects

[21]Interagency Working Group on Cyber Security and Information Assurance. 2006. *Federal Plan for Cyber Security and Information Assurance Research and Development,* Executive Office of the President of the United States, Washington, D.C., April.

[22]Thomas F. Saunders, Chair, USAF Scientific Advisory Board Summer Study, "Implications of Cyber Warfare," presentation to the committee, March 6, 2008, Washington, D.C.

[23]Defense Science Board. 2007. *Defense Science Board 2006 Summer Study on Information Management for Net-Centric Operations, Volume I, Main Report,* Office of the Under Secretary of Defense for Acquisition, Technology, and Logistics, Washington, D.C., April; presentation [of this report's results] to the committee by Vincent Vitto, Washington, D.C., March 6, 2008.

[24]National Research Council, 2007, *Toward a Safer and More Secure Cyberspace,* The National Academies Press, Washington, D.C.; and National Research Council, 2002, *Cybersecurity Today and Tomorrow: Pay Now or Pay Later,* National Academy Press, Washington, D.C.

[25]Ralph Wachter, Program Director, Software and Computer Systems, Office of Naval Research, "Overview of the Office of Naval Research and the Naval Research Laboratory's Information Assurance Related R&D," presentation to the committee, June 18, 2008, Washington, D.C.

of current and future threats. Appendix F of this report provides a representative sample and discussion of advanced security and IA concepts being pursued by academic and industrial research laboratories. This sample can serve as a starting point for the naval research community to consider as candidates for transfer into naval application.[26]

In making this suggestion to the Navy, the committee's premise is that there is no clear, demonstrably precise set of technology elements that will provide sufficient IA within the long-term vision of the GIG architecture. However, IA is an ongoing challenge that must keep ahead of the growing threats devised by clever adversaries. (It is also noted that although the research topics suggested do not include encryption and cryptography, the committee views advances in encryption as an obviously important area of IA, and at the same time it views such advances as representing only a fraction of the needed set of technologies to address fully the IA threats discussed in Chapter 1.) Overall, IA is a continuously sought end goal cutting across the entire enterprise; it must be refreshed and maintained on an ongoing basis, largely driven by (1) new research (and intelligence) into the nature of adversarial threats, (2) new concepts and techniques to counter those threats, and (3) the degree of increasing dependence on information systems to carry out critical naval missions.

A summary of suggested IA advanced research topics for the Navy is presented in Table 4.1, organized by topic under the headings Network Level, System Level, Host Level, User Level, and Privileged-User Level. (As previously noted, a more complete discussion of these topics, including current sponsoring organizations, is presented in Appendix F.) The committee suggests that these topics be used to help formulate a naval IA R&D roadmap, drawn from a yet-to-be-prioritized set of naval IA system needs. Many examples of naval IA needs are common across the Services and are spelled out in the Navy's Information Systems Security Program (ISSP).[27] Examples of naval-specific IA needs include advances to address naval afloat and marine forward-deployed forces, such as resilient networks, artificial diversity, and virtualization for security.

---

[26]Naval-specific IA needs evolve through a network-centric operation that links together Navy ships, aircraft, and shore installations into highly integrated computer and telecommunications networks, which include integrated air defense and integrating targeting data gathered by other ships and aircraft. Forward-deployed expeditionary Marines will also have specific needs associated with data integrity and availability. The DON has identified its broad-topic IA needs in its Information Systems Security Program, as summarized in Table 2.3 in Chapter 2 of this report.

[27]As described in Chapter 2, the Navy ISSP research, development, testing, and evaluation program works to provide the Navy with these essential information assurance elements: (1) assured separation of information levels and user communities, including coalition partners; (2) assurance of the telecommunications infrastructure; (3) assurance of joint user enclaves, using a defense-in-depth architecture; (4) assurance of the computing base and information store; and (5) support for assurance technologies, including a Public Key Infrastructure and directories.

TABLE 4.1 Suggested Elements of an Advanced Naval Information Assurance Research Program

Program Element	Description
Network Level	• *Border Gateway Protocol/Domain Name Service protocol "hardening"*—core network protocols responsible for routing and naming services for all Internet Protocol traffic. • *Network filtering*—filtering strategies to detect incoming attacks as well as outgoing exfiltration of sensitive information. • *Network visualization*—tools for alerting network operation to attack conditions. • *Resilient networks*—networks to ensure continual service while under denial-of-service attacks. • *Source attribution*—tools for ascertaining where a connection or attack is actually coming from. • *Decoy networking*—strategy to lure an adversary to an isolated network from which it can be monitored for intelligence (methods, behavior, and sources).
System Level	• *Secure composition*—means to ensure security properties of the whole system. • *Artificial diversity*—techniques to diversify computing fabric that allows interoperability, but also allows a change in structure of the same software for another implementation. • *Collaborative software communities*—a sharing of attack data to harden other instances of software against in-progress attacks and developing related security alert sharing technologies. • *Privacy-preserving technologies*—technologies to allow effective sharing of data while maintaining strict compartmentalization.
Host Level	• *Counter-evasion techniques for obfuscated malware*—methods to identify malware-embedded content flows. • *Virtualization for security*—technology for server consolidation and isolation of untrusted applications from the host operating system. • *Self-healing software*—software that monitors and models its own behavior. • *Hardware life-cycle tamper resistance*—techniques to detect compromises in chip-level designs and implementations during supply chain life-cycle attacks.
User Level	• *Behavior-based security*—analysis of user behavior patterns to detect threats with reasonably high reliability. • *Defense through uncertainty*—leveraging uncertainty in deployed environments to make exploitation difficult by adversary.
Privileged-User Level	• *Role- and behavior-based access control*—means of associating logical roles of a user with the specific data and applications used by the specific roles defined with an enterprise and a means of granting access to network resources. • *Self-protecting security technologies*—means of preventing denial-of-service attacks caused by a user accidentally or by design.

Current Naval Information Assurance Research and Development Budget

Based on presentations to the committee and on the Navy RDT&E (Research, Development, Test and Evaluation) Budget Item justifications documents for FY 2009, the Navy IA research budget over the past several years appears to the committee to be to grossly underfunded for properly addressing the escalating IA threats and challenges confronting the Navy. In the committee's view, this research budget appears to be underfunded even to be able to leverage the research investments of other agencies properly.[28] The information assurance basic research funding level requested by the Office of Naval Research, at approximately $2 million per year, is approximately a factor of 20 less than the yearly investment in IA research by NSF and a significant factor less than funding investments by DHS and IARPA and by other relevant agencies with science and technology programs in IA.[29] A similar deficiency is found when comparing Navy RDT&E "IA basic research" with similar DOD military service research organizations' RDT&E Exhibit R-2 budget justification. For example, the Air Force describes in its ISSP RDT&E Exhibit R-2 a requested basic research program for leveraging IA investments at DARPA, NSA, IARPA, DHS Advanced Research Activity, and leading universities.[30] The funding in this Air Force program is consistently three to four times higher than the closest similar Navy ISSP funding expressed for such activities at ONR. This Air Force program represents an example IA leveraging activity that can, in theory, provide access to leading-edge IA technology and maximum return on the Navy's IA R&D investment. The current gaps in capability for naval forces information assurance are made even more significant by a lack of strategy for investing in advanced R&D to redress these gaps, and thus should be corrected. The committee is not suggesting that the DON needs to match the investments of these other organizations, but it does need to be resourced as a

---

[28]Department of the Navy. 2008. Research, Development, Test and Evaluation, Exhibit R-2, *Fiscal Year 2009 Budget Estimates, Justification of Estimates*. Washington, D.C., February.

[29]The National Science Foundation Program Solicitation 08-521 (Cyber Trust, March 24, 2008), reported $34 million in NSF Cyber Trust program funds for FY 2008. The successor to this solicitation, NSF Program Solicitation 08-578 (CISE [Computer and Information Science and Engineering] Cross-Cutting Programs, FY 2009 and FY 2010, October 1, 2008), reports $45 million in program funds available each year in FY 2009 and FY 2010 for Trustworthy Computing. NSF, July 1, 2008, online solicitation publications: <http://nsf.gov/publications/pub_summ.jsp?ods_key=nsf08578>. Accessed April 29, 2009. DHS announced awards of $11.7 million in grants for cybersecurity research to 13 recipients from industry and academia. See *Federal Computer Week*, online publication, "DHS Awards $11.7 Million for Cyber Research," August 13, 2008. Available at <http://fcw.com/articles/2008/08/13/dhs-awards-117-million-4-cyber-research.aspx>. Accessed April 29, 2009.

[30]Department of the Air Force. 2008. "Exhibit R-2, RDT&E Budget Item Justification: 0303140F Information Systems Security Program," *Fiscal Year (FY) 2009 Budget Estimates: Research, Development, Test and Evaluation (RDT&E), Descriptive Summaries, Volume III, Budget Activity 7*, Washington, D.C., February, p. 1549. Available at <http://www.saffm.hq.af.mil/shared/media/document/AFD-080130-062.pdf>. Accessed April 29, 2009.

stronger IA R&D participant, in order to, at a minimum, understand the ongoing IA research and to be able to rapidly insert it into naval networks.[31]

MAJOR FINDING: The Department of the Navy has not established a sufficiently robust research program in IA. The funding level requested by the Office of Naval Research (ONR), approximately $2 million per year, is inadequate even to ensure that the DON effectively leverages the research investments of other agencies. Current gaps in information assurance capability for naval forces are made even more significant by a lack of strategy for investing in advanced R&D to redress these gaps.

MAJOR RECOMMENDATION: The Director, Naval Research, should develop—and the Chief of Naval Operations (CNO) and the Commandant of the Marine Corps (CMC) should ensure funding for—a robust science and technology research program in information assurance. An order-of-magnitude increase in funding levels through ONR's Naval Research Laboratory would establish the Navy as a full participant in IA technology R&D, providing the knowledge base to guide and prioritize naval implementation choices and allowing the Navy to draw from the work of outstanding members of the academic and industrial research communities. The Navy should focus its research efforts on addressing capability gaps specifically related to the needs of naval forces that are not being sufficiently addressed elsewhere.

Concurrently, the Office of Naval Research should develop a rapid technology insertion program to enable the rapid deployment of solutions for responding to new threats, based on both the leveraging of internal Navy research results and the use of ongoing research results derived from the funding of other R&D organizations, such as at the Defense Advanced Research Projects Agency, National Security Agency, Army Research Office, Air Force Office of Scientific Research, National Science Foundation, Department of Energy, and Department of Homeland Security.

[31] In deciding what approximate increase for an ONR IA R&D investment would be appropriate, the committee recognizes that information technology research is not a particularly capital-intensive undertaking; that is, the primary cost for such an investment is direct and indirect manpower cost. For example, according to R&D survey data available from the National Science Foundation (see <www.nsf.gov/sbe/srs/stats.htm>; accessed April 3, 2009), a fully funded staff-year of effort in IT R&D research effort is typically available at $200,000 to $300,000 per year, with graduate student research efforts available at a fraction of that cost. Therefore, the committee estimates that, based on a typical project staffed at three to four full-time-equivalents of effort, ONR could maintain or leverage a core group of 10 to 15 substantive research projects by increasing its current IA funding by an order of magnitude—to the neighborhood of $20 million per year. Such an increase would allow ONR to participate more broadly in the 10 priority areas for research named in the 2006 *Federal Plan for Cyber Security and Information Assurance Research and Development,* as applied to naval-specific needs (see footnote 21 above in this chapter). This suggested increase to ONR IA funding should also be accompanied by proper internal milestones and mechanisms to judge whether the investment is being managed appropriately and is yielding the expected mission benefits.

SPECIFIC CONSIDERATIONS FOR NAVAL RESEARCH AND DEVELOPMENT AND ACQUISITIONS WITH RESPECT TO INFORMATION ASSURANCE

It is generally recognized that new types of software tools that enable the exploitation of communications networks and software applications change at a very fast pace. As a result, with regard to IA tools to counter attacks, the committee places the speed of acquisition and deployment on the critical path for defense and exploitation. Correspondingly, in reviewing the R&D and acquisitions for naval forces, the committee gave significant attention to the agility that enables naval forces to conceptualize, acquire, evaluate, implement, and deploy IA technology that directly supports naval systems. Indeed, *coordination* and *integration* are the strongest enablers for agility in R&D acquisitions.

Three significant considerations for R&D and acquisitions form the basis for the committee's findings and recommendations in this area:

- Because security is a "weakest link" problem, the strength of an organization's security relies on the unified adoption of IA techniques against the most recently recognized exploits that pose significant risk. It is important to note that strong security at one end of the enterprise can be undermined by poor security in other parts of the network, and in fact by a single entity elsewhere in the network.[32]
- Time to deployment is critical. With the increasing pace of attacks, an important IA solution that is deployed too late will not be effective. The R&D organization must be agile enough to respond to threats quickly and to develop solutions that anticipate new threats.
- "Fast-time implementation processes" must incorporate the full life-cycle needs and must meet IA-related standards of implementation.[33] The processes for developing new solutions must account for (1) the time required for the establishment of funding; (2) needed research, development, testing, evaluation, and deployment time; and (3) the establishment of life-cycle support.

Existing Naval Research and Development and Acquisition Processes

A review of current acquisition legislation and management instructions provided a useful basis for the committee's finding and recommendations related to fast-time implementations. First, it appears that SECNAV Note 5000—"Rapid Development and Deployment Response to Urgent Global War on Terrorism

[32]Mark Clancy, Executive Vice President, IT Risk and Program Management, Citigroup, "Overview of Information Assurance Best Practices—A Financial Institution Perspective," presentation to the committee, July 17, 2008, Washington, D.C.

[33]The committee defines "fast-time implementation processes" as any process adaptations that are designed to deliver targeted solutions quickly and with minimal risk.

Needs"—allows the Assistant Secretary of the Navy for Research, Development and Acquisition (ASN[RDA]) "to refine the Naval Innovation Laboratory (NaIL) environment and process for rapid development and fielding of prototype solutions to meet urgent needs in the Global War on Terrorism (GWOT)."[34] To paraphrase the prescribed process, an urgent need identified from the fleet or forces can get quick consideration from a Rapid Development and Deployment Committee (RDDC). The RDDC is an ad hoc committee with membership consisting of representatives from (1) the Future Naval Capabilities Technology Oversight Group; (2) the Assistant Secretary of the Navy, Financial Management, and Comptroller (ASN[FM&C]); (3) the Deputy Chief of Naval Operations, Resources, Requirements, and Assessments (N8); (4) the Commanding General, Marine Corps Combat Development Command; and (5) the ASN(RD&A). Additional ad hoc participants are invited as needed. Figure 4.3 illustrates the fast-track process.

The catalyst for the process is an urgent need with respect to GWOT. In the committee's view, IA threats represent a concern comparable to GWOT with respect to the national security. As a result, the committee first recommends a corresponding, customized process for IA that can be initiated when an urgent need is identified.

Second, one can anticipate that the current Department of the Navy operations and maintenance (O&M) processes and controls provide an existing set of processes for addressing fast-track IA implementations as augmentation to already fielded systems, integrating the appropriate mixture of laboratory capabilities and available O&M resources. Given the general nature of anticipated IA implementations, in the current organizational structure the committee views the DON Chief Information Officer (CIO) as an appropriate office for (1) building the business cases and deciding to go forward with the needed implementations, (2) selecting the best resources for rapid implementation, and (3) setting standards and guidelines for secure technical implementation.

In addition to the committee's finding and recommendation reported below, a review of acquisition policy for the DOD C3I [command, control, communications, and intelligence] and Weapon Programs, conducted in 2007 for the Navy and the Office of the Secretary of Defense, includes an analysis of issues associated with IA acquisition.[35] The committee agrees with the findings and recommendation from that review as summarized in Box 4.1. Two of the underlying acquisition issues reported in the review were as follows: (1) There are multiple, noncoordinated policies from various authorities governing IA (no single line of authority within the Navy), and (2) for the program managers charged with

[34]Secretary of the Navy. 2008. "Rapid Development and Deployment Response to Urgent Global War on Terrorism Needs," SECNAV Notice 5000 [Cancelled SECNAVNOTE 5000, dated March 8, 2007], Department of the Navy, Washington, D.C., January, p. 1.

[35]Daniel Gonzales, Eric Landree, John Hollywood, Steven Berner, and Carolyn Wong. 2007. *Navy/OSD Collaborative Review of Acquisition Policy for DOD C3I and Weapon Programs,* RAND National Defense Research Institute, RAND Corporation, Santa Monica, Calif.

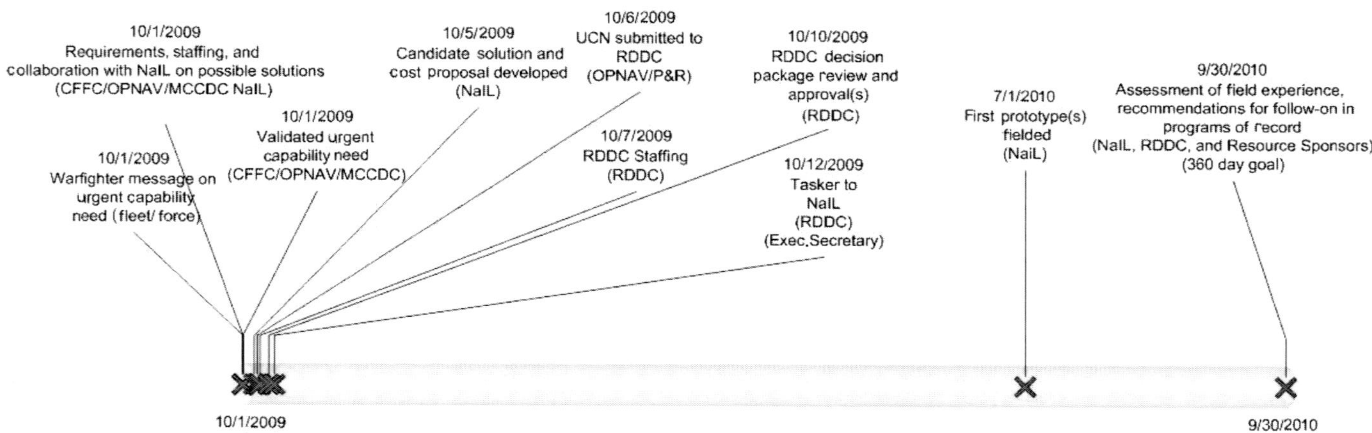

FIGURE 4.3 Time-line illustration of the rapid development and deployment (RDD) process. This figure assumes that an urgent need is determined at the beginning of FY 2010. RDD is a 1-year deployment process versus the multiyear process defined in Secretary of the Navy [SECNAV] Note 5000: "Rapid Development and Deployment Response to Urgent Global War on Terrorism Needs." NOTE: Acronyms are defined in Appendix A.

BOX 4.1
Acquisition-Related Findings and Recommendations from 2007 Navy/OSD *Collaborative Review of Acquisition Policy for DOD C3I and Weapon Program*

Finding: There has been a proliferation of IA policy documents in recent years and guidance is often not actionable.

- IA policy issuance has sharply increased in the past few years, and
- Conflicts and overlaps have been found in interoperability policy.

Finding: GIG IA guidance and standards are still evolving and not yet "stable."

- Rapid technology change,
- PA framework under development, and
- Key technologies are being developed under the leadership of industry.

Finding: Large number of IA policies—many will have to be updated to be consistent with DOD Instruction 8510.bb which is now in effect.

Recommendation: Acquisition

- Establish technology risk areas and Technology Readiness Level for GIG interoperability areas.
- Ensure independent technical assessment of GIG Program interoperability approaches by appropriate SYSCOMs or DISA.
- Move Milestone B to preliminary design review, at least for high-IT content programs.

SOURCE: Reprinted from Daniel Gonzales, Eric Landree, John Hollywood, Steven Berner, and Carolyn Wong, 2007, *Navy/OSD Collaborative Review of Acquisition Policy for DOD C3I and Weapon Programs,* RAND National Defense Research Institute, RAND Corporation, Santa Monica, Calif.

delivering complex systems, the processes for performing IA-related capability-based assessments are often separate and distinct from system acquisition decisions. This committee has independently identified the same acquisition issues for naval IA. The committee's "Organizational Considerations," this report's Chapter 6, discuss and propose potential solutions that would help mitigate these issues.

MAJOR FINDING: Cyberthreats change on a timescale much shorter than the DOD acquisition life cycle for developing and deploying cybersecurity technologies. There are increasing risks from these cyberthreats, including risks of being unable to respond to assigned warfighting missions. Rapid acquisition and fielding of IA solutions are critical, but the committee did not see processes being put into place to support this need.

MAJOR RECOMMENDATION: The committee recommends that the following specific actions be undertaken by the ASN(RDA), with the support of the Director, Naval Research, to address the timely acquisition and implementation of IA solutions:

- Actively participate in DOD efforts to define and establish intelligence that provides predictions about future cyberattack techniques which are sufficient to stimulate development of defensive responses,
- Use existing operations and maintenance processes supplemented by design and prototyping activities carried out by naval laboratories to more rapidly develop and implement solutions,
- Establish a rapid technology testing and evaluation laboratory and a technology insertion program—modeled after the Future Naval Capabilities program—to leverage and accelerate ongoing research in cybersecurity into Navy networks, and
- Establish a standard management process styled after the urgent-need process for the Global War on Terrorism (as defined in SECNAV [Secretary of the Navy] Note 5000 on "Rapid Development and Deployment Response to Urgent Global War on Terrorism Needs").

5

Application of Risk Analysis as a Basis for Prioritizing Needs

An essential problem faced by all Department of the Navy (DON) organizations with responsibilities for information assurance (IA) is how to make the inherently complex trade-offs between satisfying IA objectives and all other mission objectives. The committee believes that mission risk analysis is the appropriate foundation for IA trade-offs related to investment and system design choices. The committee has further found that the current naval efforts to apply mission risk analysis relevant to IA issues are limited and inadequate, given the magnitude of the challenge currently faced. The information assurance posture of the Navy and Marine Corps should be based on the need to maintain mission assurance at levels of risk commensurate with those accepted from other threat sources. That is almost certainly not the case today.

Risk is measured by the consequences of things that go wrong and the corresponding likelihoods of occurrence. When consequences can be extreme, the likelihood of occurrence needs to be virtually eliminated. A rigorous mission risk analysis of information assurance issues is likely to lead to a better understood and more rational set of investment and system design priorities, some of which are outlined below as recommendations. As the Navy moves to network-centric concepts of operations (CONOPS) for its fundamental missions, its overall level of mission assurance is increasingly determined by its level of information assurance and dependence. At the macro level, it is evident that electronic information system attacks can potentially provide a relatively low cost and efficient way for adversaries to reduce the effectiveness of naval warfighting capabilities. Thus, the information assurance posture and the architectural choices in DON systems should be exposed to thorough risk analysis, in the same manner that other mission-critical elements of naval systems and CONOPS are regularly exposed to mission risk analysis for more conventional threats.[1]

The committee believes that the most important area of emphasis in risk analysis in the near term should be mission-level risk created by known vulnerabilities of the entire DON network system of systems. Navy personnel interviewed and threat documents reviewed as part of the committee's deliberations indicate that the DON's current network architecture has significant vulnerabilities.[2] Although many of the vulnerabilities have mitigations and backups, there appears to be limited evaluation of the threat posed to operational missions by the threats, vulnerabilities, and mitigations as a whole. The best Navy work in the application of information risk analysis appears not to have been shared outside the organization that sponsored it. Instead, risk analysis seems to be "stove-piped," without any group forming a comprehensive picture for naval forces as a whole. As a consequence of the lack of mission-level understanding of risk, architectural choices are being made that, under certain scenarios, could even make the situation worse.

OVERVIEW AND BACKGROUND OF RISK ANALYSIS

The committee recognizes that information assurance goals and other goals for naval information systems will often be in conflict. Assurance is often expensive, and more strongly assured systems may require compromises in other areas. For example, technologies that provide opportunities for greater levels of integration and consolidation of information system functions also provide opportunities for individual exploits to have greater impact. The process of trading among competing goals is difficult because their linkages are typically complex. However, this is not the first time that such complex choices have been faced. Many other aspects of Navy system design and architecture are likewise complex and force

---

[1]General guidance and best practices on cybersecurity risk analysis and risk management for federal information technology systems are found in special publications by the Interagency Working Group, Joint Task Force Transformative Initiative, and the Computer Security Division at the National Institute of Standards and Technology (NIST). See NIST special publications (1) Gary Stoneburner, Alice Goguen, and Alexis Ferings, 2002, *Risk Management Guide for Information Technology Systems,* No. 800-30, National Institute of Standards and Technology, Gaithersburg, Md., July; (2) Ron Ross, Marianne Swanson, Gary Stoneburner, Stu Katzke, and Arnold Johnson, 2004, *Guide for the Security Certification and Accreditation of Federal Information Systems,* No. 800-37, National Institute of Standards and Technology, Gaithersburg, Md., May; and (3) Ron Ross, Stu Katzke, Arnold Johnson, Marianne Swanson, and Gary Stoneburner, 2008, *Managing Risk from Information Systems, An Organization Perspective,* No. 800-39, National Institute of Standards and Technology, Gaithersburg, Md., April.

[2]During the course of the study, the committee held discussions with a wide range of naval personnel, including not only those responsible for Navy and Marine Corps network defense, naval intelligence, and naval network architecture and system design, but also personnel responsible for network IA architecture and defense at the National Security Agency and the Defense Information Systems Agency.

trade-offs among attributes related in complex ways. Before other examples are provided showing where risk analysis in complex circumstances has been accomplished, consider the following limited subset of information assurance cases that introduce complex trade-offs between assurance and other desired system characteristics:

1. Ships are moving to consolidate their onboard networks physically onto a single, shared medium.[3] This consolidation promises to reduce costs, help reduce manning, and facilitate information sharing. However, it also introduces new IA vulnerabilities (especially with regard to denial of service) that do not exist in architectures where different networks are hosted on physically distinct communications media.
2. Resilience to denial-of-service attacks is facilitated by having a large number of Internet exchange points and by having large spare capacity. The current IA posture of the Navy, however, is to reduce the number of Internet exchange points (to facilitate monitoring), and it does not consider maintaining a greater number of exchange points for spare capacity to be cost-efficient on a day-to-day basis.
3. Guided by Department of Defense (DOD) directives, the Navy is moving strongly toward a "monoculture" of operating systems and applications by standardizing desktops. The greater the consistency of software configuration the easier it is to patch and to provide assurance against the day-to-day vulnerabilities that appear on the Internet. However, a software monoculture is also at far greater risk for catastrophic collapse induced by an attack specifically crafted to that common configuration. Is the benefit of reduced day-to-day local disruption worth the increased global risk from a sophisticated adversary?

These three examples are typical of the information assurance trade-offs requiring risk analysis at the mission level. The committee has not been presented with evidence that such analyses have been conducted comprehensively or outside of individual DON organizations.

PAST NAVY MISSION RISK ANALYSIS CONSEQUENCES

While the information assurance area may be lacking with regard to mission risk analysis, this process is not new to the Navy, and its value has been well proven. The Navy's standard practice is as follows:

- Drive system architectural choices based on desired missions capabilities,
- Recognize the threats that adversaries pose to those missions, and

[3]For example, the Navy's Consolidated Afloat Networks and Enterprise Services (CANES) program, described in Chapter 2 of this report, is designed to reduce and consolidate naval afloat network systems.

- Flow-down the identified threats, with their corresponding risks, into decisions regarding individual system developments, including both technological and operational solutions.

Consider the following non-IA examples of this standard practice:

- The Aegis system is a response to the threat of air and missile attacks on the carrier battle group. Aegis was designed to carry out a basic missile defense CONOPS, but it was also designed with numerous backup operational concepts in recognition of the likelihood that threats would partially succeed and partially degrade the system. Individual components of Aegis are designed and tested with a wide range of defined threats, both kinetic and electronic. The trade-offs involved in the design of Aegis are complicated, and include cost, operability, and impact to seaworthiness as well as mission effectiveness, but the threat and operational scenarios are the foundation for system design trade-offs.
- The existing core of secure communication systems (e.g., Military Strategic and Tactical Relay satellite and data links) was designed to mitigate the risks of known electronic attack capabilities. Those secure systems are integrated with a mission CONOPS for operating in a degraded communications environment, and for operating on the basis of objectives for—and an analysis of—achieving a mission-level capability.

What is different today is the extent and speed with which the pace of change in day-to-day operations has combined with the pace of change in information technology (IT) to lead to wholly new systems and CONOPS that have not been exposed to traditional risk analysis from an IA perspective. Navy logistics has increasingly moved to Internet applications, along with critically important joint and coalition operations support activities (e.g., U.S. Air Force tanker operations). Welfare and morale for Marines and sailors now means supporting Internet access on workstations used for day-to-day work and onboard ships. The current operational threat environment contains extensive, active, low-end attacks encountered daily, but perhaps with much higher-end (but largely invisible) threats developing in the background.

RISK ANALYSIS AND INFORMATION ASSURANCE IN THE FIELD

Risk analysis is widely, but inconsistently, used in government and industry. During its deliberations, the committee was exposed to dramatically different levels of rigor and completeness in the descriptions of how different organizations use risk analysis to drive their architectural choices.

All of the naval organizations interviewed by the committee were aware of risks and the application of risk analysis, but the degree to which they adopted risk analysis and used it to drive design and architectural choices was highly variable.

The most notable deficit is the lack of a set of common threat and operational scenarios shared across the relevant DON organizations. While combatant commands are conducting mission risk analyses for their assigned operational plans, the same scenarios were not being used by type commands and acquisition organizations. In fact, the risk scenarios that appeared to receive the greatest attention at the type commands and research organizations were quite different from the combatant command scenarios.

It might be argued that rigorous, mission-focused risk analysis is too complex at the mission level for the Navy and Marine Corps. However, the industry presenters to the committee demonstrated that quite complex analyses, and implementation of their consequences, are carried out routinely in some segments of commercial industry. An especially good example was provided by Citigroup, Inc.[4] Citigroup's information security program is driven by a regular review of all identified risk scenarios (a total of approximately 15,000 business processes that are supported by information infrastructures that expose the possibility for money losses). Information assurance investments in information infrastructure are made in response to the identified risks, and the money loss use-cases are used to verify and red-team each new application. The risk analysis approach was coupled with a set of IA principles, for example rigorously isolating every Internet-facing application. The example of Citigroup demonstrates that thorough and ongoing risk analyses can be conducted at multiple levels of abstraction (the Citigroup method extends from business processes to network design) at complexity levels comparable to those needed in the DON.

Another commercial application of risk analysis that was presented to the committee involves Verizon's use of forensic analysis to determine on a historical basis (for about 500 actual successful cyberattacks) the relative effectiveness of alternative solutions.[5] For example, the analysis presented to the committee indicated that additional improvements in system administration offered greater assurance than did additional advances in patch processing. The Navy did not have comparable efforts to understand the relative value of solutions based on actual historical case analyses.

While all of the Navy organizations that made presentations to the committee conducted some form of risk analysis, these analyses varied widely, were typically qualitative, and were limited by the scope of decision-making authority of the organization conducting the analysis. Because no substantial operations plan

---

[4]Mark Clancy, Executive Vice President, IT Risk and Program Management, Citigroup, "Information Assurance: Financial Institution Perspective," presentation to the committee, July 17, 2008, Washington, D.C.

[5]Peter Tippett, Vice President, Research and Intelligence, Verizon Security Solutions, "2008 Data Breach Investigations Report," presentation to the committee, July 18, 2008, Washington, D.C. A public copy of the report is available at <http://www.verizonbusiness.com/resources/security/databreachreport.pdf>. Accessed March 16, 2009.

(OPLAN) can be conducted within the command purview of a single command, there was a significant lack of mission-level risk analysis to guide decisions.

POSSIBLE NEW APPROACHES

While the problem of analyzing investment priorities for information assurance is difficult, it is no more complex than other Navy investment problems, although the threats and responses are prolific and less familiar to Navy leaders. As in other investment cases, threat analysis and mission scenarios (including the threat scenario) should be the foundation, and they should be coupled with metrics for mission effectiveness. Two nonexclusive approaches to mission risk analysis with respect to information systems were discussed above—operations analysis and incident analysis. In practice, both should be used. Both are reviewed briefly before discussing applications to DON missions.

Moderate extensions of conventional operations analysis methods should allow quantification of many (if not all) IA issues. The components of a conventional operations analysis model should include the following:

- An operations plan for representative conflict scenarios. Each combatant command has several such scenarios.
- A threat model, which for the purposes here should include electronic threats as well as cyberthreats to information capabilities. These should be standard threats, vetted through the naval intelligence and the IA technical communities.
- Effectiveness operational performance metrics, whose associated models should include the dependencies on network-centric capabilities.

Such mission risk analyses are carried out by a variety of DON groups. A particular example discussed with the committee was a communications risk analysis carried out by the Pacific Fleet showing the impact of reduced communications capabilities on a particular operations plan, including a variety of possible threat scenarios.[6] However, the committee did not find that these scenarios and analyses are currently being used across multiple DON organizations. While a particular combatant command may have worked out risk analyses for their OPLANs, the threat scenarios and analyses of the consequences are often not used by the type commands and others who supply critical services on which the combatant commands depend for execution. For operations analysis to be effective, it must be shared across all of the supplying and consuming stakeholders.

An alternative form of risk analysis is incident analysis and system monitoring, in which the risk to an organization's system is assessed by seeing what

[6]Robert Stephenson, Chief Technology Officer, C41 Operations, Space and Naval Warfare Command, "Maritime Communication Systems (CS) Vulnerabilities Assessment," presentation to the committee, July 18, 2008, Washington, D.C.

the organization already experiences in the threat environment in which it is immersed. Statistically based incident analysis is possible for information systems in a manner that is not feasible for platforms and weapons systems, because information systems are subjected to continuous cyberattacks; that is, while DON platforms are rarely attacked in normal operations, DON networks are continuously attacked. For incident analysis to be effective, it needs to have the following characteristics:

- It needs to be rigorous, in that incidents are followed up to discover root causes and their relationship to the efficacy of possible new preventions. In practice, it was not apparent to the committee that this is being systematically done for incidents on DON networks.
- It needs to be representative, in that the incidents tracked need to be similar to those that can potentially pose high-impact threats to mission success.

FINDINGS AND RECOMMENDATIONS

MAJOR FINDING: The Navy has not comprehensively translated adversary capabilities into risk analysis assumptions or into an operational threat, and it does not routinely share the risk analyses and threat models that exist across the various Navy and Marine Corps organizations that have responsibility for information assurance. Based on the information briefed to the committee, there does not appear to be adequate emphasis on understanding how adversaries intend to or could use their capabilities and DOD network vulnerabilities to disrupt naval operations.

As discussed previously, all of the risk analyses briefed to the committee were narrow in scope, were restricted to a single DON organization, and were not shared with other organizations with related responsibilities for information assurance. A few of the analyses did address mission risk, such as that provided by the Pacific Fleet, but they were restricted to limited threat types and did not include mission-level effectiveness metrics. The other risk analyses were technical, in that they analyzed the risks to a single system or platform but did not extend to include the mission impacts. The heart of mission risk analysis is an understanding of the adversary concept of operation and operational objectives and of the adversary capabilities available to achieve those objectives (a threat model). The DON understanding of adversaries' doctrine, CONOPS, objectives, and capabilities with respect to information assurance appears to be very limited.

MAJOR RECOMMENDATION: The Director, Naval Intelligence, in collaboration with the Defense Intelligence Agency and national intelligence organizations, should support cyber risk analysis by collecting and analyzing all source intelligence to improve the Department of the Navy's understanding of adversaries' mission intent, strategy, and tactics and to illuminate how these could impact the ability of the Navy and Marine Corps to accomplish their missions and objectives.

Some of the consequences of this recommendation are discussed in the sections that follow. An additional finding, elaborating on the original finding from the committee's letter report, concerns the need to make risk analysis not only realistic (an intelligence issue), but shared. Multiple organizations have responsibility for IA, and the information assurance capabilities likely to be needed cannot be achieved without unity of effort. To achieve unity of effort requires that the risk picture be shared.

FINDING: The extent and fast rate of change in day-to-day operations have combined with the pace of change in information technology to lead to wholly new systems and concepts of operations that have not been exposed to IA risk analysis. During its deliberations, the committee was exposed to dramatically different levels of rigor and completeness in the descriptions of how different organizations use risk analysis to drive their architectural choices. Navy and Marine Corps organizations generally recognize the importance and role of mission risk analysis, but they typically conduct such analyses only qualitatively, and the analyses are limited by the scope of each organization's decision authorities.

RECOMMENDATION: Threat and risk analysis, specifically including adversary concepts of operations and operational capabilities, should be shared across the many Navy and Marine Corps organizations with significant dependencies on information assurance. Standard scenarios and measures of effectiveness should be used by organizations responsible for information assurance.

The consequences of the recommended risk analyses should be reconciled across the Navy and Marine Corps organizations responsible for information assurance. Responsible organizations should make trade-offs related to information assurance based on the shared risk analyses.

Information Assurance Risk Considerations

Ultimately, information assurance risks to individual systems and subsystems are only relevant if they project into important mission risks—that is, if the threat can potentially prevent the Navy and Marine Corps from accomplishing their assigned missions or cause casualties during the execution of those missions. Risk analysis for information assurance should, therefore, be founded on mission risk analysis and not on risk analyses tied to individual systems or technological components.

The committee believes that a risk-based information assurance strategy, once developed from mission risk analysis, will lead to an integrated set of solutions, including the following:

• *Development of resilient systems, with the ability to "fight through" disruptions as a core design characteristic.* Just as Navy ships are designed for damage control and the ability to fight through damage, Navy networks and information systems should be designed to fight through disruptions, with graceful degradations. In furtherance of the fight-through strategy, war games should normally include risk-based disruption scenarios. Among those normally exercised should be these:

—Large-scale jamming or loss of satellite communications, eliminating dependable use of this channel of communications over various mission-critical time intervals;

—Complete denial of service to unprotected Internet/Non-Classified Internet Protocol Router Network (NIPRnet) systems over extended time intervals;

—Deceptive operations on Internet/NIPRnet-connected systems; and

—Insider-enabled attacks that either deny service or disrupt or alter information on what are otherwise protected networks.

• *A risk-based determination of the degree of isolation of various information functions so as to control the potential for attacks generated through one function to impact another integrated function.* The current naval information system environment is very large and quite complex, and comprises systems at every level of criticality from recreational functions to real-time ship and weapons control. In general, the distinctions in criticality are recognized in the defensive posture. For example, networks hosting obviously critical functions are separated either logically or physically from those hosting less-critical functions. However, the degrees of separation and the levels of monitoring applied to each need to be determined by a rigorous process of mission-by-mission risk assessment. As an example, functions of greatly differing criticality (recreation, logistics, tanker operations, and coalition communications) are currently hosted on Internet/NIPRnet-connected networks and monitored at essentially the same level. Consequently, should problems occur, the mechanisms available for partial fallbacks are very limited and nonselective.

• *Undertaking risk analyses of the mission impact of extensive denial and/or deception attacks on Internet-hosted applications.* If the mission impact of losing or compromising certain Internet-hosted applications is much greater for some applications than for others (as is almost certain), then the Navy needs to take measures to provide assurance for those applications consistent with their mission risk. Such measures might include the strict separation of nonofficial functions to alternative infrastructure (e.g., laptops and wireless local area networks), pervasive use of secure protocols among applications, concentration of monitoring on subnetworks, and sandboxing[7] of the most exposed applications.

---

[7]"Sandboxing" is a term used to describe the use of security mechanisms to isolate and control the potential spill of exploits from an untrusted program or system.

- *Extension of counterintelligence-oriented and discovery-oriented monitoring on operational Navy networks.* Current monitoring schemes strongly emphasize looking for known signatures rather than discovering previously unseen signatures. Since it is likely that high-impact attacks on Navy networks will be customized and will appear nowhere else, the Navy cannot rely solely on commercial signature databases to monitor for attacks. It needs an active discovery effort to identify attacks that may appear nowhere else.
- *More innovative and integrated approaches for bringing intelligence analysis to bear on critical near-term decisions.* Intelligence support to information assurance assessment is similar to intelligence support for other areas of threat assessment in many ways, but in important respects it is different. The committee finds that current intelligence support must be expanded to support IA needs. To elaborate on the committee's recommendation that intelligence collection and analysis must be expanded to support the impact of information assurance (and failures thereof) on mission success, the following conclusions about such intelligence collection and analysis activities are noted:
  - —Intelligence assessments must address adversarial doctrines as well as capabilities, particularly with respect to the use of enemy information operations to steal or manipulate data versus posturing for disruption in conflicts. A large naval investment in pervasive encryption could extensively thwart adversary intentions in data theft, but it would have only limited effect on preventing disruptions. Since it is unlikely that any collection can be effected that will conclusively resolve adversary intentions, the approach needed must combine collection, analysis, and examination in war games. The outcome should be one, or several, adversary concepts of operation that can be shared across the organizations responsible for information assurance.
  - —Since adversaries seek to conceal their capabilities, there is always uncertainty about these capabilities. As in other weapons system fields, estimates of capabilities must be made. The established and effective method for doing so is to invite teams of knowledgeable science and technology specialists to imagine their own approaches to the adversary concept of operation, adjusted by estimates of adversary technical capabilities. This type of activity should be conducted for threats to DON information systems, and the results should be used to prepare a threat estimate that can be shared across the organizations responsible for information assurance.
  - —An important difference between adversary information operations capabilities and kinetic capabilities is that the information capabilities are being exercised daily. The best signatures of information operations capabilities are not available through remote observation; they are on operational networks. Thus, network security and operations should be integrated with intelligence collection. Also, sensors deployed on operational networks should be selected for intelligence value and not just for their ability to pro-

vide current security. Intelligence collectors and analysts should be using the backdrop of adversary activities to improve threat models and challenge assumptions. As an example of an investigation activity that could be conducted, the committee found that little effort has been expended on estimating the scope of unobserved threats. It is impossible to measure the unobserved exactly, but a variety of methods could be used to estimate the extent of unobserved threats and bound their scope. Several examples of these methods are provided in Table 5.1.

—Information assurance, at the mission level, can be provided with a mixture of defensive and active capabilities. In principle, active and offensive methods can significantly enhance network defense, but the scenarios in which they would be effective and ineffective are not known. The committee found that no truly integrated approach to analyzing active and passive methods, or offensive and defensive methods, has been developed. Although the possibilities for synergy are real, there are also possibilities for antagonism, and there appears to be no comprehensive effort being made to disentangle the issues. In the absence of an integrated understanding, it is speculative to make investments in active or offensive methods in the hopes that the defensive posture will be improved. For active and/or offensive techniques to be incorporated into the defensive information assurance strategy, intelligence collection on cyberthreats must be greatly improved, shared, and deeply integrated into the operational plans.

—A great deal of effort is being expended in defending against and cleaning up after less sophisticated attack vectors. The volume of these attacks and their occasional effectiveness are a concern. The amount of effort expended on monitoring and cleanup of these attacks detracts from detecting more sophisticated, and likely more important, attacks. It also leads to a production-line attitude in which counts of unimportant attacks can serve to mask the lack of effort expended against sophisticated threats. The Navy should consider modifications to the Internet-exposed portions of the infrastructure that provide fundamental protections against common attack vectors (e.g., strong e-mail authentication to block many types of spear-phishing attacks).

—In developing the recommendations for organizational responses, the committee considered currently observed threats, threats that can be reasonably surmised and modeled from intelligence and technology, and unknown threats. With respect to the first two categories, it urges that risk analysis be conducted using a documented set of operational scenarios that include both known and estimated threats. With respect to unknown threats, the committee recommends ongoing science and technology research and an approach to monitoring that moves beyond the search for known signatures and uses techniques that can detect previously unknown attack vectors. With respect to each of these issues, it is of great impor-

TABLE 5.1 Example Use of Intelligence Collectors to Aid Network Defense

Example	Description
Use of honeypots[a]	• Configure honeypots behind the network's standard protection, but with solidly patched configurations, and observe whether they are compromised (or how long compromise takes). • Run a honeypot with a Web crawler configured to visit high-risk areas and observe what happens. • Forward suspicious e-mail to a honeypot with a program that automatically follows the links in the e-mail and observe whether the honeypot is compromised. • Embed some directories and files in the regular servers, placed and configured so that no legitimate user will have access to them. Observe whether the files are accessed, and if so, use careful monitoring to determine who accessed them. • Run banks of honeypots with tools to detect anything that changes, to look for zero-day exploits. Consider running banks of honeypots in this way, with different protection levels to see which are compromised or get "owned."
Use of deliberate attack	• Run attacks against the target systems that need to be defended. Originate attacks from a location outside the network using known attack vectors. If the blockage rate is less than 100 percent, then the number of known attacks can be logged directly to estimate the number of known attack types that successfully penetrate the system defenses. • Run attacks against the target systems that need to be defended. Again, originate the attacks from a location outside the network, but using methods that have not been observed being used against them previously. If the deliberate attacks get through the defenses 100 percent of the time, the system is demonstrably still vulnerable. This method will not prove whether or not these specific types of new attacks are actually occurring, but it does place bounds on the ability of the system to detect. If attacks get past system protections between zero and 100 percent of the time, the logs can be used to estimate the frequency of a specific attack.
Combined use of honeypots and controlled attacks	• Combine the use of honeypots and controlled attack to see if the controlled external attack triggers the honeypot. Ideally, the attacks will cause activity in the honeypots, even when the operational detection system does not detect them. If the attacks get past system protections between zero and 100 percent of the time, the logs can be used to estimate the frequency of a specific attack.

NOTE: The "intelligence collector" examples in this table demonstrate methodology analogous to "defect seeding" and "tag-and-release" counting methods, sometimes used in quality-assurance protocols. None of these examples provides exact count measures, but they do provide estimates based on real data. Extensive guidance on the use of honeypots is provided in cyberdefense publications, such as those found at <www.honeypots.net>. Accessed November 14, 2008.

[a]"Honeypots" are defined as "closely monitored network decoys serving several purposes: they can distract adversaries from more valuable machines on a network, they can provide early warning about new attack and exploitation trends and they allow in-depth examination of adversaries during and after exploitation of a honeypot." Definition source: <www.honeypot.net>. Accessed February 21, 2009.

tance that the results be shared across the DON and DOD organizations responsible for information assurance vectors, and that those efforts be coupled with a strong intelligence collection and analysis activity targeted at helping to improve the ability to predict future threats.

Cost Issues

The potential cost of enhanced information assurance measures and the corresponding value provided can only be assessed when risk analysis comprehensively addresses mission risk. The DON justifies very large expenditures on platforms and weapons systems precisely because their absence is estimated to place missions of great national security importance at risk. The committee believes that, increasingly, failures of information assurance will have the same large impacts on mission performance and so will justify equivalent prioritization. However, such a conclusion must be based on comprehensive evaluations of mission risk that do not yet exist.

6

Organizational Considerations

In previous chapters the committee described the challenge that cyberthreats present to the Department of the Navy's (DON's) use of network-centric operations and its dependence on commercial off-the-shelf (COTS) information technology (IT). Potential operational and technical responses that the DON might take to maintain information assurance (IA) in the face of this challenge and how it might orchestrate those responses through a risk-based management approach were also discussed.

This chapter examines potential organizational responses. It will be seen that there are many organizations, inside and outside the DON, that impact IA with respect both to the operations of naval networks and to the acquisition of naval network-based capabilities. Given this organizational complexity as well as the operational and technical complexity inherent in addressing the growing IA risks, it is recommended that the DON consider organizational realignments to better focus on the IA issues related to naval information systems and networks.

JOINT SERVICE NATURE OF INFORMATION ASSURANCE

The issues of information assurance and, more broadly, mission assurance from an information perspective for the Navy and Marine Corps are not solely Navy and Marine Corps issues. For parts of their information network infrastructure, the Navy and Marine Corps are highly dependent on joint capabilities and sometimes on systems provided by the other Services. Thus, in general, the Navy and Marine Corps will achieve mission assurance only through joint participation. Likewise, joint capabilities systems of systems are dependent on the Navy and

Marine Corps for building and operating their elements of the joint construct in ways that support the policies of the whole.

Key Trends in Cross-Service Integration

A key trend in the U.S. military is joint network-centric operations. The long-term vision is to decouple the various operational functions (e.g., sensing, targeting, weapons delivery, transport, and logistics) from individual Service platforms. A Navy ship should be able to launch a weapon on a target located by national means, provide target designation for a weapon launched by the Air Force, and draw on any Service's (or commercial) logistics stores and systems. While full network-centric capabilities are still years away, some capabilities are current and are being continually improved.

The key enabler for joint network-centric operations is information sharing. The U.S. satellite communications architecture already provides services to all Services over the same satellite links, and the Defense Information Systems Agency (DISA) provides a global communications backbone to all of the Services. Another element of cross-Service convergence is technical—namely, the increasing integration of different information service types onto fewer technical platforms. This integration is a two-edged sword. On the one hand, it leads to superior information sharing, greater efficiency, lowered costs for a given level of service, and fewer types of technical platforms to defend. On the other hand, extensive system integration could permit the possibility of losses of large-scale capabilities from single attacks. Some particular examples include the following:

- *Extensive use of commercially hosted fiber-optic and wideband satellite communications*—which has provided global broadband communications at low cost, but is significantly vulnerable to disruption and jamming;
- *Network layer convergence to everything-over-Internet Protocol (IP) and the ongoing phaseout of switched network infrastructure*—which greatly enhances network manageability and allows use of the rapidly innovating commercial IP services. However, it also opens military networks to the vulnerabilities of IP and single points of failure;[1] and
- *The convergence of unclassified and classified networks onto shared IP bandwidth enabled by cryptographic separation*—which facilitates large upgrades in bandwidth, especially for classified services; reduces the costs of providing

[1]As pointed out in the classic paper of Bellovin, the vulnerabilities of IP are intrinsic in the protocols and are not simply due to implementation issues. See Steven M. Bellovin, 1989, "Security Problems in the TCP/IP Protocol Suite," *ACM SIGCOMM Computer Communication Review*, Vol. 19, No. 3, pp. 10-19, July. See also Steven Bellovin, 2004, "A Look Back at 'Security Problems in the TCP/IP Protocol Suite,' " presented at the 20th Annual Computer Security Applications Conference, December. Available at <http://www.cs.columbia.edu/~smb/papers/ipext.pdf>. Accessed May 1, 2009.

network services by eliminating many legacy systems; and improves network manageability. However, it opens classified networks to denial-of-service attacks hosted on unclassified networks and provides an opportunity (albeit a slim one) for a compromise of the separation mechanism.

Joint Support to Navy and Marine Corps Systems

The examples above illustrate the dependence of the Navy and Marine Corps on joint systems. Without communications systems shared with the other Services (and in some cases with commercial industry and foreign partners), the capabilities of the Navy and Marine Corps would be reduced. Understanding how much they could be reduced is itself an important element of risk management and mission assurance that was highlighted earlier in this report.

The Department of Defense (DOD), as a whole, must act to ensure that plans assigned to each command are adequately supported by department-wide decisions. The Navy needs to be proactive in ensuring that plan elements assigned to the combatant Navy and Marine Corps are effectively supported in capability acquisitions. The committee finds that there are several areas where these issues are particularly evident, and there is evidence that strategies and decisions are not consistent across the whole stakeholder set.

For scenarios in which cyberattack is likely but extensive jamming and kinetic attacks are not, the most operationally effective and cost-effective approach to communications acquisition is to buy commercial fiber-optic and satellite capacity. For scenarios in which the full spectrum of threat attacks is likely, the most effective course is to acquire protected communication capabilities. The current mixed strategy being pursued by the DOD is to acquire some of each of these capabilities.

The DON must recognize the complexities inherent in pursuing the current mixed strategy. Applications that work well when high-bandwidth communications are available may not work well (or at all) in a reduced-bandwidth environment. An application and concept of operations (CONOPS) set that is designed to work well in a low-bandwidth environment must be extensively tested and exercised within that low-bandwidth environment. The operational reality might require neither the unattacked high-bandwidth services nor the secure core of low-bandwidth services, but rather a dynamically changing intermediate state. It may be that neither of the configurations that works well at either end of the service levels will work well in a dynamically changing middle ground. Moreover, the dynamically changing case is likely to be the most difficult to simulate and test.

The spectrum of potential threat environments from low to high poses a basic strategic challenge to deployed Navy and Marine Corps forces. The DON should study, in conjunction with the intelligence and research communities, whether alternative approaches to communications and application development could yield capabilities that are robustly functional across the spectrum of threat levels. This may require a partial reversal of the march toward all-COTS products, but might yield an operational system that is more robust, secure, and maintainable than the current approach of multiple fallback modes.

The DON must also strongly advocate within the joint community for the development of the capabilities that are uniquely important to Navy and Marine Corps forces. The Navy, in particular, has a dependence on mobile satellite communications that is deeper than that of the other Services. It is particularly important to the Navy that secure and protected communications capacity suitable for Navy platforms be deployed adequately for the Navy to realize the benefits of network-centric operations.

DON Support to Joint Systems

Due to the interdependence among DOD and DON systems, each Service has responsibility for keeping its own equipment and technology up to date and operational. The Joint Task Force–Global Network Operations (JTF–GNO) monitors the joint enterprise, but depends on the Services to maintain their connected systems adequately. With regard to low-sophistication cyberattacks, the updating process is central. For high-sophistication attacks, continuous patching and upgrades may yield little additional assurance. For the high-sophistication case, the DON needs an entirely different class of monitoring techniques and a science and technology (S&T)-based estimation approach, such as described in Chapter 5, to develop threat models and mitigations.

The Navy and Marine Corps are dependent on joint capabilities, but so too are those joint networks and applications dependent on the Navy and Marine Corps. If the participants in the joint network fail in their individual responsibilities, they may impact the network as a whole and the other participants. In consequence, the Navy and Marine Corps, as organizations, must consider the broader impact of their own policies and acquisitions on the health of the joint capabilities as a whole.

DOD AND DON RESPONSIBILITIES FOR INFORMATION ASSURANCE

DOD Information Assurance Responsibilities

Providing IA in the context of joint network-centric operations is the responsibility of a number of DOD organizations including the DON. The IA responsibilities of the DOD and the DON are defined in public law and in various DOD and DON instructions, directives, and memoranda.

The DOD is required to have a defense IA program under Section 2224, "Defense Information Assurance Program," of Title 10, *United States Code*. Under the provisions of the Clinger-Cohen Act of 1996,[2] the DOD is required to have a chief information officer (CIO) reporting directly to the Secretary of Defense. In DOD Directive 5144.1,[3] the Secretary has designated the Assistant Secretary of Defense for Networks and Information Integration (ASD[NII]) as the DOD CIO. DOD Directive 8500.1[4] establishes DOD IA policy and assigns organizational responsibilities. DOD Instruction 8500.2[5] provides guidance and describes procedures for implementing DOD Directive 8500.1. DOD Instruction 8580.1[6] describes how IA is integrated into the defense acquisition system.

The ASD(NII)/DOD CIO develops and promulgates IA policies, oversees appropriations for and manages the Defense Information Assurance Program (DIAP), and works with the Under Secretary of Defense for Acquisition, Technology and Logistics (USD[AT&L]) to ensure that the DOD acquisition process incorporates IA considerations consistent with the Clinger-Cohen Act requirements. The Deputy Assistant Secretary of Defense for Information and Identity Assurance (DASD[IIA]) reports to the ASD(NII) and is responsible for the DIAP and the Global Information Grid (GIG) IA portfolio, among other responsibilities. The Director of DISA assists the ASD(NII) in executing his or her responsibilities—including, in particular, the development of a single IA approach for protection of the Defense Information Systems Network (DISN).

The USD(AT&L) is tasked to ensure that IA is considered in all acquisition milestone decisions, program decision reviews, and contract awards. With the assistance and advice of the Director, Defense Research and Engineering (DDRE), the USD(AT&L) monitors and oversees IA research and technology investments, including those of the National Security Agency (NSA) and the Defense Advanced Research Projects Agency (DARPA).

The Chairman, Joint Chiefs of Staff (CJCS), provides advice and assessment of military IA capability needs and develops, coordinates, and promulgates IA policies, doctrines, and procedures for joint and combined operations.

The Commander, U.S. Strategic Command (USSTRATCOM), coordinates and directs DOD-wide computer network defense (CND) operations.

---

[2]National Defense Authorization Act for FY 1996, Public Law 104-106, formerly called the "Information Technology Management Reform Act," February 10, 1996.

[3]Department of Defense. 2005. Department of Defense Directive No. 5144.1, Washington, D.C., May 2. Available at <http://www.dtic.mil/whs/directives/corres/pdf/514401p.pdf>. Accessed May 1, 2009.

[4]Department of Defense. 2002. Department of Defense Directive No. 8500.1, Washington, D.C., October 24. Available at <http://www.niap-ccevs.org/cc-scheme/policy/dod/d85001p.pdf>. Accessed May 1, 2009.

[5]Department of Defense. 2003. Department of Defense Directive No. 8500.2, Washington, D.C., February 6. Available at <http://www.niap-ccevs.org/cc-scheme/policy/dod/d85002p.pdf>. Accessed May 1, 2009.

[6]Department of Defense. 2004. Department of Defense Directive No. 8580.1, Washington, D.C., July 9. Available at <http://www.defenselink.mil/cio-nii/docs/DoDI_8580.1pdf>. Accessed May 1, 2009.

The Director, NSA (DIRNSA), provides IA support to the DOD components, including the providing of IA and Information System Security Engineering (ISSE) services; manages the development of the IA Technical Framework (IATF); and establishes criteria and processes for evaluating and validating all IA and IA-enabled IT products used in DOD information systems. With the Director, Defense Intelligence Agency (DIA), the DIRNSA provides an IA intelligence capability. The DIRNSA is also the agent for the GIG Information Assurance Portfolio (GIAP); the GIAP management office is located at NSA and staffed with NSA and DISA personnel.

The heads of the DOD components are responsible for developing and implementing an IA program focused on DOD component-specific information and systems.

DON Information Assurance Responsibilities

Responsibilities for the IA program of the DON are defined in Secretary of the Navy Instruction 5239.3A.[7]

The DON CIO is responsible for carrying out for the Secretary of the Navy (SECNAV) the IA responsibilities assigned to the Navy by public law and by DOD directives and instructions. In particular, the DON CIO issues IA policies, integrates IA requirements with DON planning and into the DON major system acquisition management process, and serves as the focal point for IA coordination with other elements of the DOD. The DON CIO is assisted by a senior IA official (SIAO), as required by the Federal Information Security Management Act of 2002 (Public Law 107-347), and by the DON Deputy CIO (Navy) and DON Deputy CIO (Marine Corps). The Deputy CIO (Navy) is the Deputy Chief of Naval Operations for Communication Networks (OPNAV N6) and the Deputy CIO (Marine Corps) is the Director, Command, Control, Communications, and Computers.

The Assistant Secretary of the Navy for Research, Development and Acquisition (ASN[RDA]) integrates IA requirements into acquisition management of all DON IT systems and maintains an S&T program in information assurance.

The Chief of Naval Operations (CNO) develops and implements IA programs and procedures for information systems supporting Navy operations and assets, serves as the resource sponsor for Navy IA, appoints designated approving authorities (DAAs) for information systems under Navy authority, and develops Navy IA education, training, and awareness programs.

---

[7]Secretary of the Navy. 2004. SECNAV Instruction 5239.3A re: Department of the Navy Information Assurance Policy, Department of the Navy, Washington, D.C., December 20. Available at <http://doni.daps.dla.mil/Directives/05000%20General%20Management%20Security%20and%20Safety%20Services/05-200%20Management%20Program%20and%20Techniques%20Services/5239.3A.pdf>. Accessed May 1, 2009.

In Office of the Chief of Naval Operations Instruction 5239.1C,[8] the CNO assigned responsibility to OPNAV N6 for the Navy IA program, in coordination with the ASN(RDA) and the Deputy Assistant Secretary of the Navy for Command, Control, Communications, Computers and Intelligence/Electronic Warfare/Space (DASN[C4I/EW/Space]). OPNAV N6 sponsors, authorizes, and budgets for IA requirements and is instructed to "adopt an Information Technology (IT) life-cycle risk management program. . . ." The Commander, Naval Network Warfare Command (NETWARCOM), gathers and prioritizes Navy IA operational requirements from all echelon II commands. The Program Executive Office for Command, Control, Communications, Computers and Intelligence (PEO C4I) serves as the IA acquisition program manager and overall systems security engineering lead. The Director, Office of Naval Intelligence (ONI), assists OPNAV N6 and PEO C4I in the risk management process by gathering relevant threat information to assist in defining system security requirements.

The CNO has appointed the Commander, NETWARCOM, as the Navy operational DAA (ODAA) for all operating Navy collateral/General Services (GENSER) information systems, networks, and telecommunications systems and has assigned the Navy echelon II commanders as the developmental DAAs.[9] He has appointed the Commander, Space and Naval Warfare Systems Command (SPAWAR), as the Navy certification authority for collateral/GENSER classified and unclassified, information, telecommunications, and network systems.

Other important responsibilities of the Commander, NETWARCOM, as defined in Office of the Chief of Naval Operations Instruction 5239.1C include computer network vulnerability testing and providing training to fleet units. As discussed below, NETWARCOM also has an operational role in conducting and directing CND.

The Commandant of the Marine Corps (CMC) has IA responsibilities parallel to those of the CNO.

The process by which naval IA policies are translated into system capabilities is illustrated in Figure 6.1. A DON program manager receives IA policy guidance from a number of sources, including the FORCEnet Enterprise Architecture, the DOD IT Standards Registry (DISR), and the GIG IA Technical Framework (GIATF). As indicated above, a number of DOD and Navy organizations are involved in setting these policies.

Each program's ISSE activity is responsible for discovering users' information protection needs and then designing and making information systems to safely resist the threats to which the program may be subjected. According to

---

[8]Chief of Naval Operations. 2008. OPNAV Instruction 5239.1C., Department of the Navy, Washington, D.C., August 20. Available at <http://www.fas.org/irp/doddir/navy/opnavinst/5239_1c.pdf>. Accessed May 1, 2009.

[9]OPNAV 89 was appointed as the DAA for special access programs, and the Director, ONI, as the Navy liaison to the NSA DAA for all sensitive compartmented information (SCI) program systems.

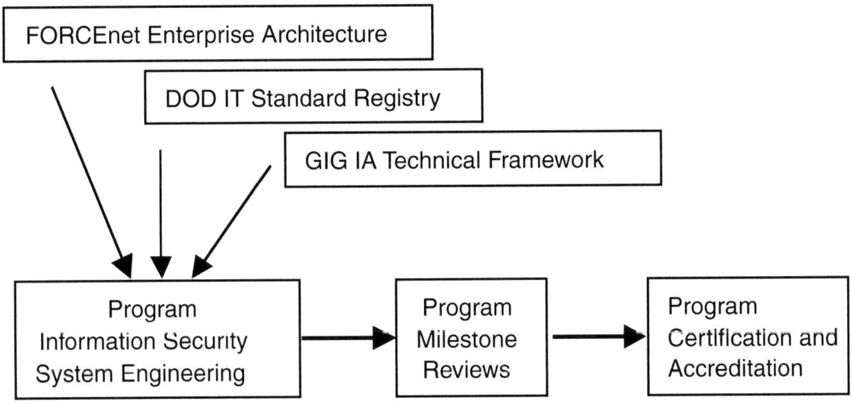

FIGURE 6.1 Process for information assurance (IA) policies translation into the Department of the Navy system capabilities. NOTE: Acronyms are defined in Appendix A.

DOD Instruction 8580.1, for any acquisitions of Automated Information Systems (AIS), outsourced IT-based processes, and platforms or weapon systems with IT interconnections to the GIG, the program manager needs to appoint an IA manager. The IA manager determines the system mission assurance category (MAC) and confidentiality level, identifies and implements appropriate system baseline IA controls, and plans and executes the certification and accreditation (C&A) process. For acquisitions that are designated as "mission-critical" or "mission-essential" systems, the IA manager must also prepare and submit an acquisition IA strategy.[10]

Acquisition IA strategies for all acquisition category (ACAT) IAM, ACAT IAC, and ACAT ID programs[11] must be approved by the DOD component CIO and submitted to the DOD CIO for review prior to all acquisition milestone decisions, program decision reviews, and acquisition contract awards. The heads of the DOD components are delegated the authority to conduct reviews of acquisition IA strategies on behalf of the DOD CIO for all other acquisitions, and may delegate authority to approve acquisition IA strategies.

[10]DOD Instruction 8580.1 provides definitions and guidance for "mission essential" and "mission critical" designations for IT systems. Such designations must be made by a Component Head, a Combatant Commander, or their designee. Available at <http://www.defenselink.mil/cio-nii/docs/DoDI_85801.pdf>. Accessed February 11, 2009.

[11]Acquisition Category (ACAT) I programs are major defense acquisition programs. For ACAT ID programs, the USD(AT&L) is the Milestone Decision Authority (the "D" in "ID" refers to the Defense Acquisition Board). For ACAT IAC programs, the head of the DOD component is the Milestone Decision Authority (the "C" in "IAC" refers to the Component CIO). For ACAT IAM programs, the ASD(NII)/DOD CIO is the Milestone Decision Authority (the "M" in "IAM" refers to the Major Automated Information Systems Review Council).

The PEO C4I and the Navy PEO for Enterprise Information Systems (EIS) under the ASN(RDA) manage most programs involving IT. However, PEO Ships (e.g., DDG-1000, LPD 17 [landing platform dock]) and PEO Aircraft Carriers (e.g., CVN-76 [nuclear-powered aircraft carrier]), and the Marine Corps Systems Command manage several programs in which computing and networking infrastructure are being procured along with ships. The program management offices for the PEO C4I and the PEO EIS are staffed largely with personnel drawn from SPAWAR.

According to Secretary of the Navy Instruction 5400.15C, the commanders of Naval Air Systems Command (NAVAIR), Naval Sea Systems Command (NAVSEA), SPAWAR, and Marine Corps Systems Command (MARCORSYSCOM) exercise technical authority (TA)[12] and certification authority for weapons and IT systems. In particular, program managers must obtain certification from SPAWAR or MARCORSYSCOM that a weapon and/or information system being developed has satisfied information assurance requirements. As mentioned above, operational system accreditation resides with the Commander, NETWARCOM, as the ODAA.

From an operational perspective, at the DOD level, USSTRATCOM has been assigned responsibility for coordinating and directing CND. The JTF–GNO is the USSTRATCOM element that implements this responsibility. The DISA commander is dual-hatted as the JTF–GNO commander. Navy CND is the responsibility of the NETWARCOM and of its subordinate element, the Navy Cyber Defense Operations Command (NCDOC), which is the Navy CND service provider. The Marine Corps network defense falls to the Marine Corps Network Operations and Security Center (MCNOSC).

From the above descriptions, it is apparent that numerous DOD and DON organizations are involved in IA. These organizations are endeavoring to work collaboratively, and have developed various forums such as the Naval NETWAR FORCEnet Enterprise (NNFE)[13] and the Cyber Asset Reduction and Security (CARS) Task Force to facilitate this collaboration. Nevertheless, the committee is concerned that there is too great an opportunity for debilitating delays in responding to IA problems and for critical errors in responding to IA problems—both due to seams in the process of developing IA policy, developing requirements for IA, funding the acquisition of IA capabilities, developing and acquiring systems requiring IA, and operating these systems.

The next section addresses more centralized organizational options for the Navy to consider in order to avoid these seams. (See Table 6.1 for a summary of current Department of the Navy information assurance responsibilities.)

---

[12]Technical authority is the authority, responsibility, and accountability to establish, monitor, and approve technical standards, tools, and processes in conformance with applicable DOD and DON policy, requirements, architectures, and standards.

[13]The NNFE focuses on command, control, communications, computers, combat systems, and intelligence (C5I) systems and appropriate business IT solutions. It is chaired by the Commander, NETWARCOM, acting as the chief executive officer; the Commander, SPAWAR, acts as the chief operations officer, and OPNAV N6 acts as the chief financial officer.

the committee believes that the DON will need to have a cadre of officers, enlisted personnel, government civilians, and contractors who can responsibly integrate COTS components into sensitive network-centric warfare applications, with sufficient attention to IA so as to manage the trade-offs between IA and mission performance in system design and mission operations. The IA management team must develop the strategies to cope with changing disruptive threats during the life subcycles of design, development, and field support.

The committee's opinion is that the department is not structured to accomplish this objective effectively today. There currently exist multiple stakeholders, including acquisition authorities, resource sponsors, systems commands, PEOs, and operational commands, with varying authorities relative to achieving information superiority. This structure results in the knowledge, authority, and accountability being very broadly dispersed—in the committee's view, too broadly dispersed to deal with the recognized complexities associated with IA in a timely manner, with time controlled by ever-changing adversarial capabilities.

Architectural Alignment

Information assurance today suffers from a "traditional" and overly limiting definition of practice. Information assurance is more than simply ensuring proper password practices, guarding against network intrusions by installing firewalls, and providing patch updates when required. In its broadest sense, IA can be described as the absolutely essential, always ongoing process, involving people, procedures, and technology, required to protect a highly networked naval force against attacks to its communications capabilities and the data therein. A successful cyberattack will put critical DON data and information in jeopardy and thus potentially reduce the capability of the DON to execute its missions. In that sense, the committee, through its deliberations, assesses that there are multiple seams across the information assurance area in the DON that might prevent the development and execution of a unified, integrated information assurance strategy. Coordination among policy, acquisition, financial resource allocation, operations, and manpower and training functions and authorities is greatly complicated by these seams. As an example and consequence, the synchronization of software architecture, hardware architecture, and organizational design/enterprise architectures either does not routinely occur or is accomplished with difficulty. This results in the lack of an authoritative information assurance architecture that is adequately scoped and programmed and in a lack of configuration control relative to information assurance. It also does not easily permit adjustments related to unanticipated changes in threat, potentially rendering newly developed capabilities as higher-than-desired risk elements for the naval forces structure. This misalignment can exist within the DON and across the agencies and the other military Services.

As with many other system attributes, information assurance cannot be "installed" at system testing. The needs and requirements for information assur-

readiness that the department requires. Suggestions and specific recommendations for organizational integration that has promise to achieve more effective information assurance are offered.

For purposes of clarity and precision, the term "networks" is used in what follows to refer to large general-purpose or enterprise systems such as the Navy/Marine Corps Intranet (NMCI), the Marine Corps Enterprise Network (MCEN), shipboard local area networks (LANs), aviation general-purpose networks, and DOD networks such as the Non-Classified Internet Protocol Router Network (NIPRnet) and the Secure Internet Protocol Router Network (SIPRnet) and so on, used for command, control, and intelligence purposes. But the term "networks" is not used to refer to combat system networks such as the Joint Tactical Information Distribution System, Multi-functional Information Distribution System, or Cooperative Engagement Capability. In a Venn diagram of networks and applications, applications are included in the network set only for considerations of hosting, transport, and policies relative to degradation of performance. In practice, a network designated approving authority would accredit the use of a certified application to use a network. However, the network authority would not get involved with the application's function—that is the purview of the application's process owner.

Although the committee was briefed and saw evidence on the convergence of combat system command and control with intelligence networks, its deliberations were premised on the continued separation of these networks in naval warfare.

Mention is also made of "life-cycle information assurance." By this term, the committee is referring to the need to provide information assurance capability throughout the life cycle of a system. This especially becomes significant when a system transitions from the acquisition community to the operating forces and is subjected to operations and maintenance (O&M) resource pressures.

The discussion below articulates the reasons why information assurance is critically important for future naval warfighting success—and, correspondingly, why the Department of the Navy needs to place its development and management in the hands of a dedicated cadre, provided with appropriate educational and training support.

Intellectual Property

The DON does not currently own or control the designs of the critical technology components that comprise the information capabilities designed and operated as part of the network-centric command-and-control systems. However, the DON does design how commercial off-the-shelf components are integrated and used to achieve desired warfighting and system capabilities. The use of COTS components offers significant economic and performance advantages, but they come with inherent IA risks outlined and discussed in previous chapters. In order to respond to the high level of IA risk associated with a COTS component strategy,

TABLE 6.1 Continued

Functional Area	Organization	Responsibilities
Acquisition	ASN(RDA)	Oversee acquisition of all DON IA capabilities and ensure compliance.
	OPNAV N6	Draft and maintain Navy's IA acquisition master plan; coordinate fleet requirements for acquisition of communications security.
	PEO C4I	Manage the Navy's IA acquisition programs and projects, including R&D and full life-cycle support.
	PEOs	Oversee program acquisition execution in area of jurisdiction.
	SYSCOMs	Oversee program acquisition execution in area of jurisdiction.
	MARCORSYSCOM	Procure USMC IA programs.
	DISA	Direct the procurement of DOD-wide IA products and licenses.
Certification and accreditation	SPAWAR	Serve as Navy's certification authority for information and network systems.
	NETWARCOM	Serve as Navy's accreditation authority for information and network systems.
	PEOs	Apply IA architectures and IA requirements in program execution.
	SYSCOMs	Integrate IA requirements in design of information systems.
	MARCORSYSCOM	Serve as USMC certification and accreditation authority for systems.
	HQMC	Serve as USMC certification and accreditation authority for networks.

NOTE: Acronyms are defined in Appendix A.
SOURCE: Derived from Office of the Chief of Naval Operations Instruction 5239.1C, Department of Defense Instruction 8500.2, and Department of Defense Instruction 8580.1.

INTEGRATED POLICY DEVELOPMENT AND ORGANIZATIONAL SUPPORT

The previous chapters of this report offer the background and context in which information assurance should be viewed by the DON for today's and tomorrow's warfighting environment. The subsections in this final major section of Chapter 6 illuminate IA policies and processes as currently addressed and implemented, and identify weaknesses in achieving the necessary IA posture and

TABLE 6.1 Current Naval Information Assurance (IA) Responsibilities

Functional Area	Organization	Responsibilities
Operational requirements	OPNAV N6	Assure overall IA program execution in coordination with ASN(RDA) and DASN C4I; sponsor, authorize, and budget for IA requirements.
	NETWARCOM	Serve as Navy Computer Network Defense (CND) Service Provider and coordinate defense of Navy computer networks as directed by JTF–GNO; provide CND training to fleet units as requested by fleet commanders; prioritize Navy IA operational requirements via input from Echelon II commands.
	OPNAV N89	Computer Network Defense Service Provider for special access systems.
	MCCDC/HQMC	Identify USMC IA requirements and capabilities.
	JTF–GNO	Direct and coordinate the defense of all DOD computer networks.
	DISA	Establish connection requirements and approval for the Defense Information Systems Network.
	ONI	Provide threat input and IA risk management assistance to OPNAV N6 and PEO C4I.
Policy	DON CIO/DASN C4I	Provide overall DON IA policy guidance and focal point for IA; coordination with other elements of the DOD.
	OPNAV N6/HQMC	Approve and issue IA policy, systems management, and metrics documents for Navy and USMC.
	NETWARCOM	Provide guidance for implementation of Navy C&A policy; write safeguarding and accounting policies for DON COMSEC materials.
Manpower and training	OPNAV N6	Oversee Navy IA training requirements and provide requirements to the Personnel and Training and Standing Team (PTST).
	OPNAV N1	Develop Navy schoolhouse IA training and education; ensure that IA training is incorporated into pertinent Navy training and appropriate formal schools.
	NETWARCOM	Manage the DON communication security training program.
	PTST	Identify Navy IA billet and establish IA training requirements for military and civilian personnel.
	HQMC/MCCDC	Develop USMC IA training, manpower, and education requirements.

continued

TABLE 6.1 Current Naval Information Assurance (IA) Responsibilities

Functional Area	Organization	Responsibilities
Operational requirements	OPNAV N6	Assure overall IA program execution in coordination with ASN(RDA) and DASN C4I; sponsor, authorize, and budget for IA requirements.
	NETWARCOM	Serve as Navy Computer Network Defense (CND) Service Provider and coordinate defense of Navy computer networks as directed by JTF–GNO; provide CND training to fleet units as requested by fleet commanders; prioritize Navy IA operational requirements via input from Echelon II commands.
	OPNAV N89	Computer Network Defense Service Provider for special access systems.
	MCCDC/HQMC	Identify USMC IA requirements and capabilities.
	JTF–GNO	Direct and coordinate the defense of all DOD computer networks.
	DISA	Establish connection requirements and approval for the Defense Information Systems Network.
	ONI	Provide threat input and IA risk management assistance to OPNAV N6 and PEO C4I.
Policy	DON CIO/DASN C4I	Provide overall DON IA policy guidance and focal point for IA; coordination with other elements of the DOD.
	OPNAV N6/HQMC	Approve and issue IA policy, systems management, and metrics documents for Navy and USMC.
	NETWARCOM	Provide guidance for implementation of Navy C&A policy; write safeguarding and accounting policies for DON COMSEC materials.
Manpower and training	OPNAV N6	Oversee Navy IA training requirements and provide requirements to the Personnel and Training and Standing Team (PTST).
	OPNAV N1	Develop Navy schoolhouse IA training and education; ensure that IA training is incorporated into pertinent Navy training and appropriate formal schools.
	NETWARCOM	Manage the DON communication security training program.
	PTST	Identify Navy IA billet and establish IA training requirements for military and civilian personnel.
	HQMC/MCCDC	Develop USMC IA training, manpower, and education requirements.

continued

TABLE 6.1 Continued

Functional Area	Organization	Responsibilities
Acquisition	ASN(RDA)	Oversee acquisition of all DON IA capabilities and ensure compliance.
	OPNAV N6	Draft and maintain Navy's IA acquisition master plan; coordinate fleet requirements for acquisition of communications security.
	PEO C4I	Manage the Navy's IA acquisition programs and projects, including R&D and full life-cycle support.
	PEOs	Oversee program acquisition execution in area of jurisdiction.
	SYSCOMs	Oversee program acquisition execution in area of jurisdiction.
	MARCORSYSCOM	Procure USMC IA programs.
	DISA	Direct the procurement of DOD-wide IA products and licenses.
Certification and accreditation	SPAWAR	Serve as Navy's certification authority for information and network systems.
	NETWARCOM	Serve as Navy's accreditation authority for information and network systems.
	PEOs	Apply IA architectures and IA requirements in program execution.
	SYSCOMs	Integrate IA requirements in design of information systems.
	MARCORSYSCOM	Serve as USMC certification and accreditation authority for systems.
	HQMC	Serve as USMC certification and accreditation authority for networks.

NOTE: Acronyms are defined in Appendix A.
SOURCE: Derived from Office of the Chief of Naval Operations Instruction 5239.1C, Department of Defense Instruction 8500.2, and Department of Defense Instruction 8580.1.

INTEGRATED POLICY DEVELOPMENT AND ORGANIZATIONAL SUPPORT

The previous chapters of this report offer the background and context in which information assurance should be viewed by the DON for today's and tomorrow's warfighting environment. The subsections in this final major section of Chapter 6 illuminate IA policies and processes as currently addressed and implemented, and identify weaknesses in achieving the necessary IA posture and

ance and its effect on the hardware, software, and the operational environment must be continuously considered as a system is being designed and developed. Information assurance, including the potential need for adjustments due to changing threats, has to be considered in the early design decisions and trade-offs at the front end of the life cycle or it will be very difficult and costly to deal with later during development or in operation.

Outsourcing and Acquisition

The acquisition of major naval networks from industry is clear recognition that the intellectual property for these networks does not wholly reside in the DON or DOD. Moreover, the lack of a fully authoritative and effective DON information assurance CONOPS and information assurance enterprise architectures complicates major network acquisitions such as the NMCI, the DDG-1000 Total Ship Computing Environment, the LPD 17 Shipboard Wide Area Network, the USS *Ronald Reagan* CVN 76 Integrated Communication Advanced Network, and the Littoral Combat Ship network platforms, including both its hardware and software. Potential implications include the following:

- *The life-cycle information assurance and the required strong configuration control handoff from systems commands and PEOs to the fleet degrading over time.* Operational and resource pressures can negatively impact IA system upgrades and personnel training and can create challenges to life-cycle configuration management;
- *The Navy's losing the capability to understand or effectively manage network-centric technology processes owing to the dispersion of know-how regarding threats, system IA architecture and design, system development, and system field operations.* The lack of a dedicated, coherent "network workforce" community at all of the systems commands and in the fleet amplifies this trend; and
- *The Navy's not fully integrating contractors into its operational processes although it has outsourced much of its required technological capabilities to industry.* Of even greater concern is how much second- and third-level outsourcing has occurred, resulting in additional vendors and correspondingly reduced visibility.

Organizational Structure

Structurally complicating the complete elimination of the IA seams resulting from differing policies, requirements, financial resource allocations, acquisitions, operations, and manpower and training functions is that there are two military Services within the DON. This necessarily involves consideration of multiple and varied requirements affecting network-centric operations. Priority differences within the Navy and the Marine Corps can often yield different results for network-centric capability, which often must be reconciled at the SECNAV level or go unresolved. For policy and acquisition issues, this is typically accomplished by two different organizations within the department that address network-centric, IA issues: (1) the DON CIO organization and (2) the ASN(RDA) organization.

A depiction of the many entities involved in the IA process for the DON was presented previously (see Table 6.1). The committee assesses the information assurance governance in the department to be too complicated and far less than optimal. In fact, as a National Research Council committee concluded in 2000, "Currently no single individual within the Department of the Navy has IA governance responsibility and authority."[14] This remains the case today.

Need for Organizational Realignment

The reasons cited in this chapter, combined with the above cited description of information assurance governance in the DON, suggest the need for organizational realignment. The department should examine alternatives to acquiring and managing networks that provide tightly controlled IA discipline with respect to architecture conformance, life-cycle support, and configuration management; an ability to accommodate technology insertion; and a structure to facilitate risk management. In an effort to gain insight into organizational models that might help to accomplish these objectives, the committee examined the Naval Nuclear Propulsion Program (NNPP) and the Department of the Army Chief Information Officer/Assistant Chief of Staff for Information Management (DOA CIO/G6) organization.

Naval Nuclear Propulsion Program

The Naval Nuclear Propulsion Program is known for its effective management and accountability for safety assurance. For example, after the Columbia Space Shuttle accident in 2003, the Director, NNPP, was called to testify before Congress on the NNPP and its culture of safety "that has allowed Naval Reactors to be successful for the last 55 years."[15] More recently, the Director, NNPP, was assigned by the Secretary of Defense to investigate the mistaken delivery by the U.S. Air Force of fuses used in intercontinental ballistic missiles to Taiwan.[16]

---

[14]Naval Studies Board, National Research Council. 2000. *Network Centric Naval Forces: A Transition Strategy for Enhancing Operational Capabilities,* National Academy Press, Washington, D.C., pp. 217-218.

[15]Statement of Admiral F.L. "Skip" Bowman, USN, Director, Naval Nuclear Propulsion Program, before the House Committee on Science, Washington, D.C., October 29, 2003. See also, NNBE Benchmarking Team, 2003, *NASA/Navy Benchmarking Exchange (NNBE)*, Vol. II, Progress Report, Naval Sea Systems Command and National Aeronautics and Space Administration, Washington, D.C., July 15.

[16]Secretary of Defense Task Force on DoD Nuclear Weapons Management. 2008. *Report of the Secretary of Defense Task Force on DOD Nuclear Weapons Management, Phase I: The Air Force's Nuclear Mission,* Washington, D.C., September.

Under Executive Order 12344,[17] the NNPP was established as a program carried out by the DON and the Department of Energy (DOE) and led by a director with technical background and experience in naval nuclear propulsion who serves for a term of 8 years.[18] The director, if a Navy officer (all directors have been Navy officers so far), is an admiral reporting directly to the CNO and having direct access to the Secretary of the Navy, and is also an assistant secretary of the DOE. The NNPP has total responsibility for all aspects of Navy nuclear propulsion, including research, design, construction, testing, operation, maintenance, and ultimate disposition of naval nuclear propulsion plants; the safety of reactors, including the prescribing and enforcement of standards and regulations; personnel, including training and concurrence in the selection of all personnel who operate reactors; and administration, including oversight of procurement, logistics, and fiscal management.

The NNPP and the position of Director, NNPP, are certainly unique aspects of the Navy management structure, in response to the high-priority need for specialization and safety accountability. Because of the authorities granted the director, the potential seams between policy, requirements, budgeting, research, acquisition, operations, and training and personnel management that the committee observed for IA and networking are not present for nuclear reactors. The committee understands that there are significant differences between providing reactors and networks for ships and that the governance structure of the NNPP is due to unique factors—including the legacy of Admiral Hyman Rickover, USN—that could not simply be replicated for IA and networks.

Nonetheless, the committee believes that there are strong parallels between the nuclear propulsion area and the IA area. In the analysis of the committee, the parallels—which include the need for strong alignment of authorities and responsibilities; the need for strong leadership and continuity (in the case of the NNPP, facilitated by the qualifications, high rank, and long tenure of the director); the emphasis on selection and training of technically qualified personnel; and the need for strong, continuing technical support (in the case of the NNPP, provided by the DOE laboratories)—call for a similar organizational response. A takeaway lesson from the NNPP model is that there is a clear and strong sense of ownership of the nuclear propulsion mission and the applicable authorities.

Department of the Army Chief Information Officer

The DOA CIO/G-6 provides architecture, governance, portfolio management, strategy, C4 IT acquisition oversight, and operational capabilities to enable joint

[17]Ronald Reagan, President of the United States. 1982. *Executive Order 12344* (Naval Nuclear Propulsion Program), The White House, Washington, D.C., February 1.

[18]The program is also known as the Naval Sea Systems Command (NAVSEA) Nuclear Propulsion Directorate (08), or NAVSEA 08.

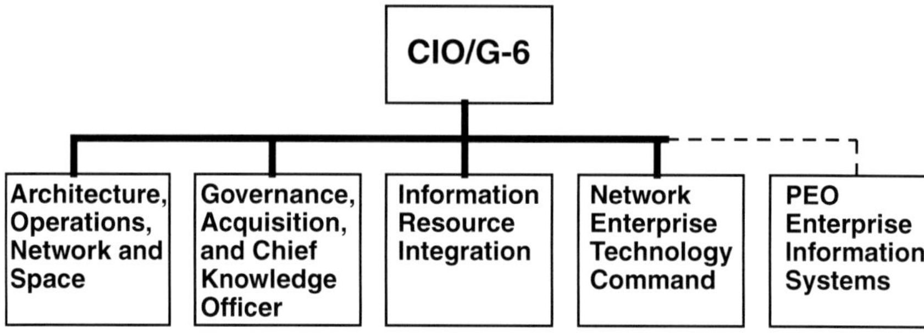

FIGURE 6.2 Organizational structure of the Department of the Army, Chief Information Office (CIO)/G-6.

network-centric operations for the Army.[19] The DOA CIO/G-6 is a lieutenant general who reports to the Secretary of the Army and provides staff support to the Chief of Staff of the Army. The DOA CIO/G-6 organization is depicted in Figure 6.2.

The Army's Network Enterprise Technology Command (NETCOM) reports directly to the DOA CIO/G-6 and operates and defends LandWarNet—the Army's portion of the GIG. The NETCOM commander is a major general. Composed of more than 17,000 soldiers, civilians, and contractors, the signal commands and brigades of NETCOM are stationed and deployed worldwide, supporting Army, joint, interagency, and multinational operations, and the Pentagon.

The PEO EIS develops, acquires, integrates, deploys, and sustains network-centric information technology, business management, communications, and infrastructure systems. PEO EIS reports on a solid-line basis directly to the Assistant Secretary of the Army for Acquisition, Logistics and Technology (ASA[ALT]) and on a dotted-line basis to the DOA CIO/G-6.

The DOA CIO/G-6 is the principal focal point for the Army for information management matters with external organizations; it has authority over policy, requirements, budgeting, operations, and training and personnel management; it is the DAA for Army information systems[20] (with the exception of Army sensitive compartmented information [SCI] systems); and supports the ASA(ALT) acquisition of information systems and parts of other major capabilities. While the mission, organization, and culture of the Department of the Army are not the same as those of the Department of the Navy—in particular, as noted above, the

[19]Headquarters, Department of the Army. 2008. "Army Knowledge Management and Information Technology, Army Regulation 25-1," Washington, D.C., December 4.

[20]The CIO/G-6 may delegate the DAA role. The Army certification authority (CA) is the Army senior IA officer. The Director, Office of Information Assurance and Compliance (an element of NETCOM), has been appointed as the SIAO by the DOA CIO/G-6. The CA maintains a list of qualified government organizations to perform the certification activities.

DON has two Services—the DOA CIO/G-6 organization provides an example of an alternative organization and governance structure for networks and IA, with the DOA CIO/G-6 explicitly tasked to fill the seams between the various Army organizations with a specific role regarding IA and networking.

Alternative Organizational Models

Based on its analysis of the current organizational structure of the Navy for networking and IA[21] and after consideration of models such as the NNPP and the DOA CIO/G-6, the committee considered various new alternative organizational constructs. In anticipation that the DON may contemplate the inclusion of an organizational response in its efforts to address the information assurance challenges outlined in this report, the committee has developed four naval IA organizational model alternatives, which are presented below.

The most comprehensive organizational approach is considered in Option 1; Options 2 through 4 would involve somewhat less change. A chart depicts the structure of each. Elements of these options could also be selectively implemented.

IA Organizational Model—Option 1

Option 1 (Figure 6.3) would establish a new senior flag/general officer position, entitled Director, Naval Networks (DNN), to rotate between the Navy and Marine Corps, as the single authority for naval networks. The DNN would provide the strong leadership that is needed for secure operation of naval networks in a similar fashion to the strong leadership provided by the Director, Naval Reactors, for the secure operation of naval reactors. A uniformed officer is preferred over a civilian to emphasize clearly the operational importance of the position. This dual-reporting position would assume the current functions of the DON CIO and the DASN(C4I/EW/Space) and would report directly to the Secretary of the Navy, for acquisition oversight of naval network systems and fulfilling Clinger-Cohen Act responsibilities.[22] The position would also report to the CNO and the CMC with responsibility for life-cycle management of information systems afloat and

[21]The committee believes that management of IA cannot be separated from management of networking, in the meaning of the term defined earlier in this chapter. Therefore, its organizational recommendations cover both IA and networking.

[22]The Clinger-Cohen Act of 1996 (Public Law 104-106) outlines the requirements for acquisition of information technologies in government agencies and the responsibilities of the agency chief information officer. Any DON IA organizational adaptation must also conform to the requirements of the Goldwater-Nichols DOD Reorganization Act of 1986 (Public Law 99-433). Under this act, the Secretary of the Navy has explicit authority to assign such of his powers, functions, and duties as he considers appropriate to the Under Secretary of the Navy and to the Assistant Secretaries. The Secretary of the Navy has made the ASN(RDA) responsible to "establish policy and procedures and manage all research, development and acquisition" within the department (Public Law 99-433, Section 5015).

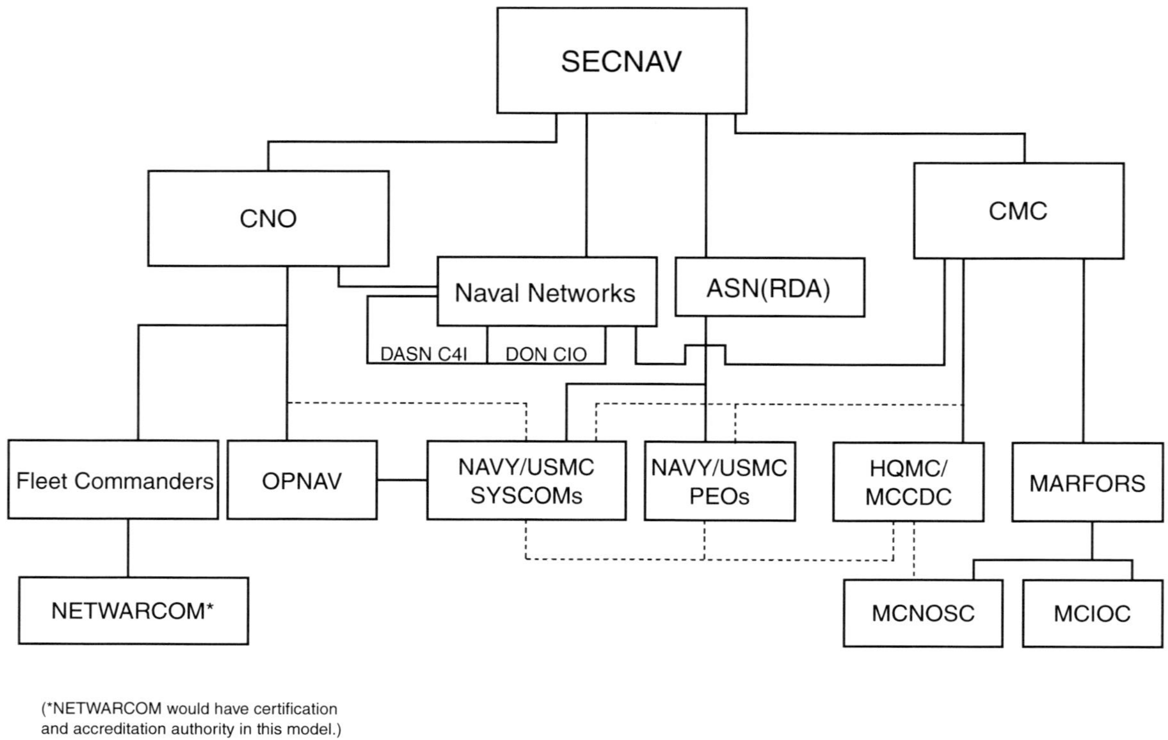

FIGURE 6.3 Organizational model—Option 1: Adding "Naval Networks" organization (senior flag or general officer with triple hat) to the Secretary of the Navy (SECNAV), the Chief of Naval Operations (CNO), and the Commandant of the Marine Corps (CMC). NOTE: MARFORS, Marine forces. Other acronyms are defined in Appendix A.

ashore and for the education and training of a dedicated officer, enlisted, and civilian cyber workforce. The office would be appointive for at least the duration of the Program Objective Memorandum (POM) cycle (5 years) to ensure policy and execution continuity and accountability.

The DNN would have dotted-line relationships with OPNAV N6 and Headquarters, Marine Corps (HQMC) for requirements and resource issues; with NETWARCOM, MCNOSC, and the Marine Corps Information Operation Center for operational issues; and with the ASN(RDA) for acquisition issues. The DNN would also be responsible for integrating IA strategies and plans across all naval communities (surface, subsurface, expeditionary, air, space, and cyberspace), as well as with joint communities. The Director, Naval Networks, would have the authority to establish network "safe-to-operate" criteria to use as enforcement authority if a naval network was judged to be so impaired as to potentially harm naval operations.[23]

This model would retain the Naval Network Warfare Command (NETWARCOM) at the Echelon III level as the functional and operational type commander for Navy networks, but would also grant NETWARCOM and HQMC C4 the authority to certify as well as accredit software and hardware systems on naval networks. This alternative would consolidate significant responsibility for IA policy, acquisition, financial resource allocation, operations, and manpower and training functions under the DNN.

Establishing the position of DNN would recognize the critical importance of networking to current and future naval capabilities. It would also represent a historic step comparable to the establishment of the NNPP.

The committee believes that acquiring network capability for the DON and providing the necessary life-cycle support and the needed education and training must be executed at the highest levels within the department to achieve the right organizational response. The DNN would also be given post-program, post-budget adjustment authority to accommodate exigencies that might occur during the development, production, and fielding of information and network systems, specifically to coordinate IA capabilities. This organizational alignment would afford great benefits by merging the DON CIO and the DASN C4I responsibilities. It would permit the DNN to employ both Clinger-Cohen Act and DOD Directive 5000 acquisition directives to optimal benefit for the DON. The combination of the offices would also bridge the transition of networks from the acquisition domain into the operating forces by the office's reporting to SECNAV and to the CNO and CMC. This would give the Director, Naval Networks, the responsibility to ensure life-cycle support of networks.

---

[23]Such a "safe-to-operate" decision may involve the important operational risk analysis of "network gain/loss" versus "operational gain/loss." That is, leaving a network connected could allow an intrusion to propagate, but disconnecting the network could cause the failure of a mission and possible loss of life if the mission was dependent on network connectivity.

Like the Director, Naval Reactors, the DNN would have to have ready access to technical expertise to provide technical depth and continuity of knowledge. This would be provided by SPAWAR and the Navy laboratories, augmented as necessary by support from Federally-Funded Research and Development Centers and contractors.

Sole execution authority within the DON would be given to NETWARCOM and HQMC C4 to both certify and accredit information systems, thus centralizing authority for this most critical IA requirement. NETWARCOM would designate certification authorities and establish independent verification and validation teams for periodically and frequently checking approved certifications in both the acquisition and operational stages. The DNN would coordinate with naval operational and intelligence agencies to develop cyberthreat analyses.

Due to the DNN's stature, tenure in office, and technical support, the DNN would be well positioned to address other key issues identified in this report, including energizing the Navy's research program in IA and CND, integrating offensive and defensive cyber operations, and integrating all aspects of IA through a risk management approach.

IA Organizational Model—Option 2

Option 2 (Figure 6.4) would establish a Network Programs Office (NPO) as a Direct Reporting Program Manager (DRPM) reporting to the ASN(RDA), transferring or adding required support resources as needed from the Navy's PEO C4I and PEO EIS and appropriate USMC PEOs, to ensure a high level of attention to challenging acquisitions and strict acquisition discipline for the delivery of afloat and ashore networks and for their life-cycle management and information assurance readiness.

In this model, NETWARCOM is retained at the Echelon III level as the functional and operational type commander for Navy networks; likewise, MCNOSC retains its current authorities and responsibilities in the Marine Corps. As in Option 1, this option would also grant NETWARCOM and HQMC C4 the sole authority to certify as well as accredit software and hardware systems on naval networks. This alternative therefore modifies naval IA policy and acquisition only. It does not change financial resource allocation, operations, or manpower and training functions.

The establishment of the Network Programs Office as a Direct Reporting Program Manager would provide the special scrutiny and oversight necessary for significant, challenging new acquisitions in the network domain. Sole authority within the DON is given to NETWARCOM and HQMC C4 both to certify and accredit information systems, thus centralizing authority for this most critical IA requirement.

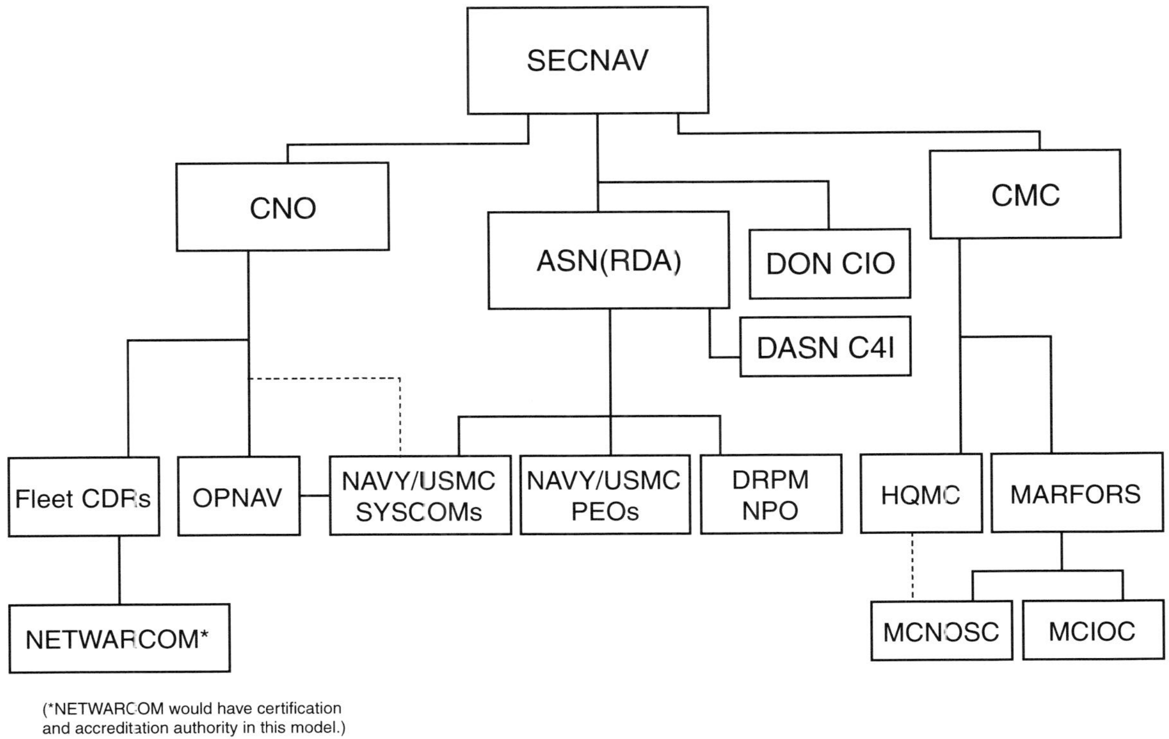

FIGURE 6.4 Information assurance organizational model—Option 2: Adding a "Network Programs Office" (NPO) as a Direct Reporting Program Manager (DRPM) reporting to ASN(RDA). NOTE: CDR, commander; MARFORS, Marine forces. Other acronyms are defined in Appendix A.

IA Organizational Model—Option 3

Option 3 would elevate NETWARCOM to the Echelon II level reporting to the CNO, thus recognizing the Navy-wide criticality of information assurance and networks (Figure 6.5). This model also grants NETWARCOM and HQMC C4 the sole authority to certify as well as to accredit software and hardware systems on naval networks. This alternative modifies policy and potentially financial resource allocation, and also modifies manpower and training functions. It does not change acquisition and operations.

Placing NETWARCOM as an Echelon II command would recognize the Navy-wide importance of information assurance and make this important function report directly to the CNO. Establishing NETWARCOM as an Echelon II command would give NETWARCOM the clear enforcement responsibility for network IA policy and operations across the entire Navy enterprise. Increased influence with OPNAV in the Program Planning and Budgeting System process would result, as NETWARCOM will provide information and network requirements directly to the OPNAV staff. As in Options 1 and 2, sole authority within the DON would be given to NETWARCOM and HQMC C4 both to certify and to accredit information systems, thus centralizing authority for this most critical information assurance requirement.

IA Organizational Model—Option 4

The committee's Option 4 model represents the least amount of change with respect to current naval IA operations. This option would grant NETWARCOM and HQMC C4 the sole authority to certify as well as to accredit software and hardware systems on naval networks (Figure 6.6). Thus, this alternative would only modify naval IA policy responsibilities. It would not change acquisition, financial resource allocation, operations, and manpower and training functions. (See Table 6.2 for a summary comparison of each option discussed above.)

Summary Discussion

The committee consulted with several senior naval officials who were selected on the basis of their potentially providing the committee with new insights concerning possible organizational recommendations. These officials included the current Director of Naval Nuclear Propulsion, the former ASN(RDA), and the current Commander of NETWARCOM. They were also chosen to help the committee understand issues associated with the currently "federated" approach for governing naval IA and addressing IA issues. On the basis of its discussion with the selected officials, the committee's own analysis and experienced-based personal views, and the Navy's projections regarding the growing threats to information assurance, the committee believes that Option 1

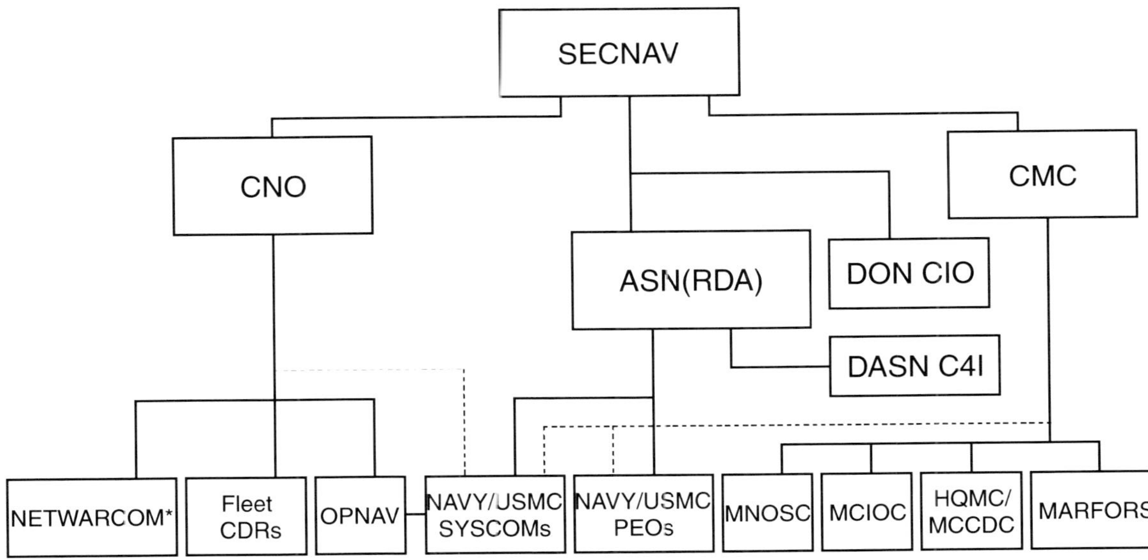

(*NETWARCOM would have certification and accreditation authority in this model.)

FIGURE 6.5 Information assurance organizational model—Option 3: The Naval Network Warfare Command (NETWARCOM) with additional information assurance authorities at the Echelon II level. NOTE: CDR, commander; MARFORS, Marine forces. Other acronyms are defined in Appendix A.

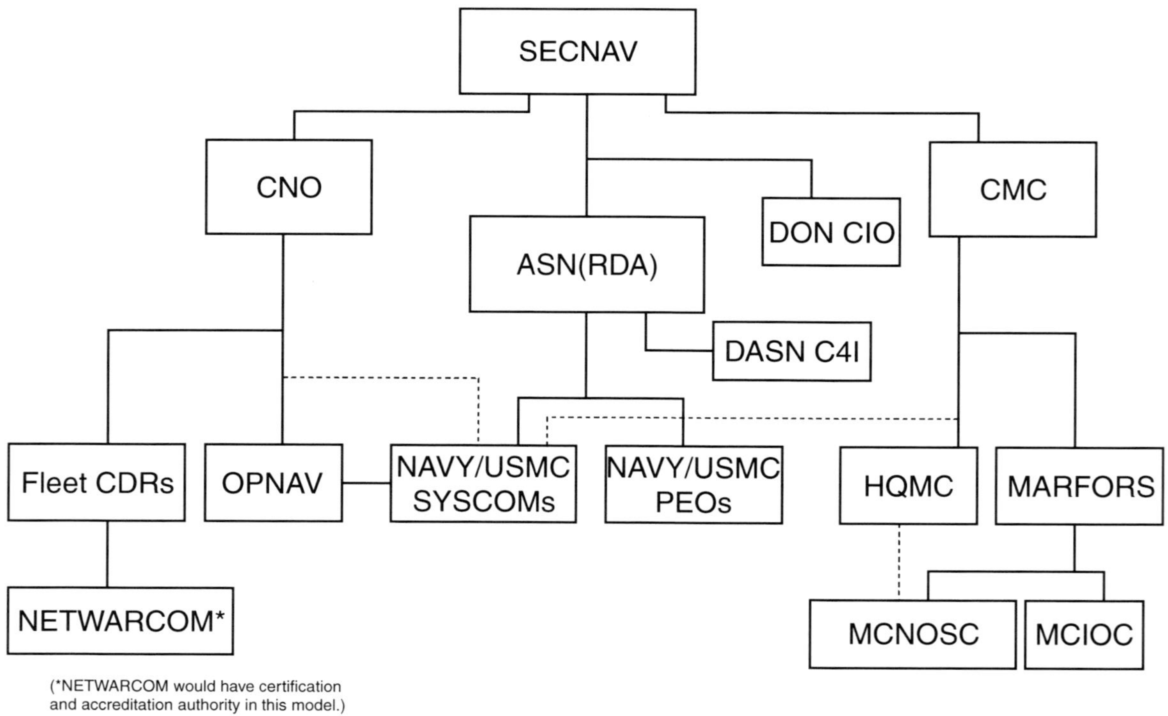

FIGURE 6.6 Information assurance organizational model—Option 4: The Naval Network Warfare Command (NETWARCOM) and the Marine Corps Network Operations and Security Command (MCNOSC) with additional information assurance authorities. NOTE: CDR, commander; MARFORS, Marine forces. Other acronyms are defined in Appendix A.

TABLE 6.2 Comparison of Alternative Organizational Model Constructs and Their Impact on Naval Information Assurance Functional Areas

Proposed Organizational Construct	Naval Information Assurance Functional Area Impacted				
	Policy	Acquisition	Resource Allocation	Operations	Manpower and Training
Naval Networks (Option 1)	Combine DON CIO and DASN (C4I/EW/Space) Sole authority for C&A to NETWARCOM, HQMC C4	Combine DON CIO and DASN C4I	Combine DON CIO and DASN C4I Coordinate with OPNAV N6	Safe to operate Adds cyberthreat analysis to NETWARCOM, MCNOSC	Directs naval networks manpower and training Manager, cyber workforce
Direct Reporting Network Programs Office (Option 2)	Sole authority for C&A to NETWARCOM, HQMC C4	DRPM ASN(RDA)	No change	Adds cyberthreat analysis to NETWARCOM, MCNOSC	No change
NETWARCOM Echelon II (Option 3)	Sole authority for C&A to NETWARCOM, HQMC C4	No change	Program Objective Memorandum (POM) Major actor	Adds cyberthreat analysis to NETWARCOM, MCNOSC	Directs naval networks manpower and training
NETWARCOM, HQMC C4 Additional IA Authorities (Option 4)	Sole authority for C&A to NETWARCOM, HQMC C4	No change	No change	Adds cyberthreat analysis to NETWARCOM, MCNOSC	No change

NOTE: Acronyms are defined in Appendix A.

would best position the Navy and the Marine Corps to address current and future information assurance and cyber-related challenges and to facilitate rapid IA progress. The committee does not suggest that the models described above are exhaustive in their scope; of the four organizational models presented, however, Option 1 provides the most clear and comprehensive naval IA governance authority and responsibility for addressing the IA issues outlined throughout this report, including the previously discussed governance seams between naval IA functions of policy, acquisition, financial resource allocation, operations, and manpower and training. The Option 1 model provides a clear and strong signal for the ownership and accountability of the bedrock DON information assurance mission.

With the appropriate assignment of authority and responsibility to the Director of Naval Networks, Option 1 would more closely resemble the clear cyber command lines of authority and responsibility found in the Headquarters (HQ) U.S. Army.[24] The Army HQ's model for managing cyber-related activities is in contrast to the current DON federated approach for managing naval IA and networking, and would appear to provide the opportunity for clearer governance responsibilities and cleaner, unambiguous lines of authority.[25] By providing a single focal point for naval cyber matters, the proposed naval Option 1 construct would also facilitate relationships with joint organizations, ensuring that the DON speaks with a single voice.

As a less dramatic potential naval IA organizational approach, a "strong federated" governance model—an option in which each of multiple parties has well-defined responsibilities with a clear understanding of the relations among those responsibilities, an improvement over the current "weak federated" model—is also recognized by the committee to provide a partial solution to naval IA governance issues. However, a federated approach lacks clear accountability for many crosscutting IA and network operations-related issues, and it leaves unreconciled potentially critical IA issues such as (1) the need for fast response and decision making in the time of crisis, (2) the development and continuity of deep knowledge and properly trained manpower in crosscutting cyber technical areas, (3) the ongoing requirement for IA resource prioritization with different organizational points of impact, and (4) the development of required expertise to manage and balance more systematically the high-level IA-related trade-offs and operational risks.

---

[24]See Army Regulation 25-1, Headquarters, Department of the Army, Washington D.C., December 4, 2008; and Capt Carla Pampe, USAF, 8th Air Force Public Affairs Office, 2006, "Air Force Officials Consolidate Network Ops," Department of the Air Force, Barksdale Air Force Base, La., July. Available at <http://www.af.mil/news/story.asp?id=123023090>. Accessed May 1, 2009.

[25]Note, however, that Army cyber field support operations are distributed between NETCOM and the Intelligence and Security Command (including its subordinate element, the 1st Information Operations Command [Land]), whereas the Navy's cyber operations are consolidated under NETWARCOM.

As with any suggested organizational model, the preferred centralized command model for naval IA presented by the committee will have disadvantages as well as advantages. For example, centralized organizational structures are sometimes viewed as less innovative, and perhaps less adaptive, than structures in which multiple or occasional competing authorities coexist. Also, less centralized structures are typically better at horizontal and multiple-direction communication than are the centralized structures, which are sometimes dominated by hierarchical, top-to-bottom communications.

Nonetheless, the committee's opinion is that the organizational structure required to address the four potentially critical IA issues just listed, coupled with the growing cyberthreat and the resulting need for clear IA accountability, point to Option 1 as the preferred model. While Options 2, 3, and 4 are less-extensive variations of the theme expressed in Option 1, the committee's opinion is that IA and related network operations will demand more clear governance authority and single-line accountability than are provided by Options 2, 3, and 4, especially as network-centric operations, information assurance, and cyberwarfare all grow in importance over the coming years.[26]

A DON decision and potential implementation of Option 1, or of any model outlined above, would obviously require further in-depth study and deliberation. However, the urgency of addressing information assurance and cyberdefense needs calls for a new organizational model on which serious examination should begin immediately. The committee recognizes that an organizational change to the recommended Option 1 would be a major step for the DON; however, the committee also believes that, as suggested by one senior Navy leader, such a change is better achieved through the vision and drive of a determined group of naval leaders than in response to a major cyber-related catastrophic event.

MAJOR FINDING: The governance of information assurance is widely distributed across naval forces, with many parties playing roles, resulting in many governance seams. In particular, there is no centralized authority or organizational mechanism in place in the Department of the Navy for governing IA and end-to-end cyber operations. For example, a shared scope of governance of security policy and fiscal authority for naval networks resides throughout the DON, including with the Department of the Navy Chief Information Officer; the Deputy CNO for Network Operations; Headquarters, Marine Corps; Naval Network Warfare Command; Echelon II Chief Information Officers; Commander–Naval Installation Command; Program Executive Officers; and Navy Systems Command.

[26]For example, a significant finding from the investigation of recent errors involving the mistaken shipment of nuclear weapons by the U.S. Air Force was the lack of clear lines of authority, which allowed safety assurance practices to degrade over the years. In other words, "no one owned the problem." ADM Kirkland Donald, USN, Director, Naval Nuclear Propulsion, private communication with committee co-chairs, October 10, 2008.

MAJOR RECOMMENDATION: The leadership of the Department of the Navy should examine more-centralized IA-related organizational structures for integrating its information assurance strategies and plans across all naval communities (surface, subsurface, expeditionary, air, space, and cyberspace), as well as for integrating those same strategies and plans with joint communities (Combatant Command, Office of the Secretary of Defense). The examination should address the needed IA governance and fiscal authorities for sustaining both current and future readiness levels, as well as which DON organizations are critical to defending against evolving cyberthreats—from the strategic to the tactical level.

Appendixes

Appendix A

Acronyms and Abbreviations

ABNCP	Airborne Command Post
ACAT	acquisition category
ADNS	Advanced Digital Network System
AFOSR	Air Force Office of Scientific Research
AFSAB	Air Force Scientific Advisory Board
AIS	Automated Information Systems; Automatic Identification System
AMF	Airborne, Maritime, Fixed Station
ARO	Army Research Office
ASA(ALT)	Assistant Secretary of the Army for Acquisition, Logistics, and Technology
ASD(NII)	Assistant Secretary of Defense for Networks and Information Integration
ASN(FM&C)	Assistant Secretary of the Navy, Financial Management and Comptroller
ASN(RDA)	Assistant Secretary of the Navy for Research, Development and Acquisition

NOTE: Key terms used in this report are consistent with the definitions of terms provided in the U.S. government's "National Information Assurance (IA) Glossary," CNSS Instruction 4009, Revised June 2006, published by the Committee on National Security Systems; and as provided in DOD Directive 8500.2—"Information Assurance Implementation." These documents are available at <http://www.cnss.gov/Assets/pdf/cnssi_4009.pdf> (accessed February 4, 2009) and <http://www.niap-ccevs.org/cc-scheme/policy/dod/d85002p.pdf> (accessed February 4, 2009), respectively. Any unique key terms used in the report and not specifically addressed in the official IA glossary are defined in the report when first used, and included in this appendix.

ATM	asynchronous transfer mode
BGP	Border Gateway Protocol
C&A	certification and accreditation
C2	command and control
C3I	command, control, communications, and intelligence
C4	command, control, communications, and computers
C4I	command, control, communications, computers, and intelligence
C4ISR	command, control, communications, computers, intelligence, surveillance, and reconnaissance
CAC	common access card
CANES	Consolidated Afloat Networks and Enterprise Services
CARS	Cyber Asset Reduction and Security
CCA	Clinger-Cohen Act of 1996
CCE	common computing environment
CDL	common data link
CENTRIXS	Combined Enterprise Regional Information Exchange System
CFFC	Commander, U.S. Fleet Forces Command
CI	counterintelligence
CIO	chief information officer
CJCS	Chairman, Joint Chiefs of Staff
CMC	Commandant of the Marine Corps
CNA	computer network attack
CNCI	Comprehensive National Cybersecurity Initiative
CNCS	Centralized Net Control Station
CND	computer network defense
CNE	computer network exploitation
CNO	Chief of Naval Operations
COCOM	combatant commander
COMSEC	communications security
CONOPS	concepts of operations
COTS	commercial off-the-shelf (includes commercial open-source software)
CT	cryptologic technician
CTN	cryptologic technician, networks
DAA	designated approving authority
DAR	data at rest
DARPA	Defense Advanced Research Projects Agency

DASD(IIA)	Deputy Assistant Secretary of Defense for Information and Identity Assurance
DASN	Deputy Assistant Secretary of the Navy
DASN(C4I/EW/Space)	Deputy Assistant Secretary of the Navy for Command, Control, Communications, Computers and Intelligence/Electronic Warfare/Space
DASN(IIA)	Deputy Assistant Secretary of the Navy for Information and Identity Assurance
DCGS	Distributed Common Ground System
DCIO	Deputy Chief Information Officer
DDoS	distributed denial of service
DDRE	Director, Defense Research and Engineering
DHS	Department of Homeland Security
DIA	Defense Intelligence Agency
DIAP	Defense Information Assurance Program
DIRECT	Defense Injection Reception EAM Command and Control (C2) Terminals
DIRNSA	Director, NSA
DISA	Defense Information Systems Agency
DISN	Defense Information Systems Network
DISR	Department of Defense Information Technology Standards Registry
DNN	Director, Naval Networks
DNS	Domain Name System
DOA CIO/G6	Department of the Army, Chief Information Officer/ Assistant Chief of Staff for Information Management
DOD	Department of Defense
DOE	Department of Energy
DON	Department of the Navy
DRPM	Direct Reporting Program Manager
DSB	Defense Science Board
DWTS	Digital Wideband Transmission System
EA	enterprise architecture
EAM	emergency action message
EIS	Enterprise Information Systems
EMI	electromagnetic interference
EMP	electromagnetic pulse
EPLRS	Enhanced Position Location Reporting System
EW	electronic warfare
FNC	Future Naval Capabilities
FSBS	Fixed Submarine Broadcast Site

FTP	File Transfer Protocol
FW	firewall
G6	Communications Electronics Division (USMC)
GCCS-M	Global Command and Control System-Maritime
GCS	Global Communications System
GENSER	General Services
GIAP	GIG Information Assurance Portfolio
GIATF	GIG Information Assurance Technical Framework
GIG	Global Information Grid
GNOSC	Global Network and Operations Security Center
GPS	Global Positioning System
GWOT	Global War on Terrorism
HBSS	Host Based Security System
HF	high frequency
HIDS	host-based intrusion detection system
HQMC	Headquarters, Marine Corps
IA	information assurance
IARPA	Intelligence Advanced Research Projects Activity
IASM	Intelligent Agent Security Manager
IATF	Information Assurance Technical Framework
IC	intelligence community
IM	instant messaging
IMAP	Internet Message Access Protocol
IO	instructor/operator
IP	Internet Protocol
IPv6	Internet Protocol version 6
ISNS	Integrated Shipboard Network System
ISO	International Organization for Standardization
ISR	intelligence, surveillance, and reconnaissance
ISSE	Information System Security Engineering
ISSP	Information Systems Security Program
IT	information technology; information systems technician
IT-21	IT for the 21st Century
JFCC-NW	Joint Functional Component Command–for Network Warfare
JMSDF	Japan Maritime Self-Defense Force
JTF–GNO	Joint Task Force–Global Network Operations
JTRS	Joint Tactical Radio System

JWICS	Joint Worldwide Intelligence Communications System
LANT	Atlantic Fleet
MAC	mission assurance category
MAGTF	Marine Air-Ground Task Force
MARCORSYSCOM	Marine Corps Systems Command
MCCDC	Marine Corps Combat Development Command
MCEITS	Marines Corps Enterprise IT Services
MCEN	Marine Corps Enterprise Network
MCI	Marine Corps Installation
MCIOC	Marine Corps Information Operation Center
MCNOSC	Marine Corps Network Operations and Security Command
MDA	Maritime Domain Awareness; Milestone Decision Authority
MHQ/MOC	Maritime Headquarters, Maritime Operations Center
MIDS	Multi-functional Information Distribution System
MILSTAR	Military Strategic and Tactical Relay (satellite)
MITSC	Marine Information Technology Support Center
MUOS	Mobile User Objective System
N6	Deputy Chief of Naval Operations for Communications Networks
NaIL	Naval Innovation Laboratory
NAOC	National Airborne Operations Center
NAVAIRSYSCOM	Naval Air Systems Command
NAVNETCOM	Naval Network Command
NAVSEA	Naval Sea Systems Command
NCDOC	Navy Cyber Defense Operations Command
NCES	Net-Centric Enterprise Service
NCIS	Naval Criminal Investigative Service
NETCOM	Network Enterprise Technology Command (Army)
NETOPS	Network Operations
NETWAR	Network Warfare
NETWARCOM	Naval Network Warfare Command
NGEN	Next Generation Enterprise Network
NIPRnet	Non-Classified Internet Protocol Router Network
NITSC	Naval Information Technology Support Center
NMCI	Navy Marine Corps Intranet
NNFE	Naval NETWAR FORCEnet Enterprise
NNPP	Naval Nuclear Propulsion Program
NOC	Network Operations Center

NPO	Network Programs Office
NRC	National Research Counsil
NRL	Naval Research Laboratory
NSA	National Security Agency
NSB	Naval Studies Board
NSF	National Science Foundation
NWS	Naval Warfighting Systems
O&M	operations and maintainance
OACE	Open Architecture Computing Environment
OAET	Open Architecture Enterprise Team
ODAA	operational designated approval authority
ODASD(I&IA)	Office of the Deputy Assistant Secretary of Defense for Information and Identity Assurance
ODBC	open database connectivity
ONE-Net	Overseas Navy Enterprise Network
ONI	Office of Naval Intelligence
ONR	Office of Naval Research
OPLAN	operations plan
OPNAV	Office of the Chief of Naval Operations
OPNAV N1	Deputy Chief of Naval Operations (Navy Total Force)
OPNAV N6	Deputy Chief of Naval Operations for Communications Networks
OSD	Office of the Secretary of Defense
P&R	personnel and readiness
P2P	peer-to-peer
PA	performance allocation
PAC	Pacific Fleet
PACOM	Pacific Command
PEO C4I	Program Executive Office for Command, Control, Communications, Computers and Intelligence
PEO IWS	Program Executive Office for Integrated Warfare Systems
PHP	Hypertext Preprocessor
PII	Personal Identifiable Information
PKI	Public Key Infrastructure
PLA	People's Liberation Army
PM	program manager
POM	Program Objective Memorandum
POP	post office protocol; Point of Presence (USMC)
PRC	People's Republic of China
PTST	Personnel and Training and Standing Team

R&D	research and development
RDDC	Rapid Development and Deployment Committee
RDT&E	research, development, testing and evaluation
RNOSC	Regional Network and Operations Center
S&T	science and technology
SATCOM	satellite communications
SCCVI	Secure Configuration Compliance Validation Initiative
SCI	sensitive compartmented information
SCRI	Secure Configuration Remediation Initiative
SECNAV	Secretary of the Navy
SIAO	senior information assurance official
SIPRnet	Secret Internet Protocol Router Network
SMCC	Survivable Mobile Command Center
SMTP	Simple Mail Transfer Protocol
SOA	service-oriented architecture
SPAWAR	Space and Naval Warfare Systems
SPAWARSYSCOM	Space and Naval Warfare Systems Command
SSBN	ballistic missile submarine
STEP	Standardized Tactical Entry Point (program)
SYSCOM	Systems Command
TA	technical authority
TCP	Transmission Control Protocol
TDL	tactical data link
TDN	tactical data network
TF	technical framework
TG	Transformation Group
TOG	Technology Oversight Group
TS	Top Secret
TSAT	Transformational Satellite Communications
TTPs	tactics, techniques and procedures
UCN	Urgent Capability Need
USAF	United States Air Force
USB	universal serial bus
USD(AT&L)	Under Secretary of Defense for Acquisition, Technology and Logistics
USMC	United States Marine Corps
USN	United States Navy
USSTRATCOM	United States Strategic Command
USW DSS	Undersea Warfare Decision Support System

VLAN	virtual local area network
VOIP	Voice over Internet Protocol
VPN	virtual private network
VSCAN	virus scan
WIFI	wireless fidelity
XSS	cross-site scripting

Appendix B

Terms of Reference

At the request of the Chief of Naval Operations, the Naval Studies Board of the National Academies will conduct a study to examine information assurance for network-centric naval forces. Specifically, the study will:

- Review the Department of Defense and the Department of the Navy responsibilities for information assurance, to include policies, plans, and manuals, and identify competing and non-competing areas of responsibility between the Departments and within the Department of the Navy, as well as recommend any organizational adaptations which facilitate rapid progress;
- Review recent information assurance-related studies conducted by and for the Department of Defense and Department of the Navy, and summarize their key recommendations and implementation status;
- Examine the Department of Defense and Department of Navy research, development, and acquisition process for information assurance, and recommend alternative approaches to the process that allow for greater flexibility and response time in meeting the information assurance requirements of network-centric naval forces;
- Assess potential information assurance vulnerabilities for network-centric naval forces, to include the "last mile" of information passed to embarked forces, and identify the appropriate technology and operational means to mitigate their vulnerabilities when operating only with U.S. military forces, or coalition forces;
- Identify methodologies, including experimentation, for dealing with degraded performance and the loss of warfighting system integrity, particularly important to the effectiveness of network-centric naval forces, due to a lack of information assurance;

- Review and recommend information assurance best practices from critical industrial and commercial operations applicable to the Department of Navy and its FORCEnet initiatives;
- Assess the role of different information architecture constructs, including information assurance approaches, for managing risks (e.g., building specially-protected "sub-nets" to handle particularly sensitive, high consequence information); and
- Recommend investment analysis approaches, excluding cost as a consideration, for managing cyber attack risks to network-centric naval forces that address the consequences of possible cyber attacks, the likelihoods of these attacks actually occurring, and the uncertainties surrounding assumptions about these risks.

This 12-month study will produce two reports: (1) a letter report following the second full committee meeting that summarizes the key information assurance initiatives underway within the Naval NETWAR/FORCEnet Enterprise and recommends any near-term information assurance needs for network-centric naval forces, to include any defense-related efforts that the naval forces should take advantage of and/or assure compatibility with; and (2) a comprehensive report that addresses the full terms of reference.

Appendix C

Biographies of Committee Members

Barry M. Horowitz (NAE), *Co-Chair*, is professor of systems engineering at the University of Virginia. His areas of expertise include the design and development of large-scale networks and information systems; application of security technology to large, network-based commercial systems; and the design of large systems that involve coupling private data systems or mission-critical support systems with open networks, such as the Internet. He previously served as chair and founder of Concept Five Technologies and as president and chief executive officer of the MITRE Corporation and of Mitretek Systems. He has served on numerous scientific boards and advisory committees, including as a member of the National Research Council's (NRC's) Committee on Freight Transportation Information Systems Security. Dr. Horowitz is a member of the National Academy of Engineering and is a current member of the Naval Studies Board.

Nils R. Sandell, Jr., *Co-Chair*, is vice president and general manager of BAE Systems Advanced Information Technologies. His areas of expertise include military command, control, communications, computers, intelligence, surveillance, and reconnaissance (C4ISR) systems and technologies. He is a former associate professor at the Massachusetts Institute of Technology (MIT), where he lectured in estimation and control theory, stochastic processes, and computer systems. Dr. Sandell has served on numerous scientific boards and advisory committees, including as a member of the NRC Committee on Network-Centric Naval Forces and co-chair of the NRC Committee on C4ISR for Future Naval Strike Groups. Currently, he is a member of the NRC Committee on Operational Science and Technology Options for Defeating Improvised Explosive Devices, and he served on the NRC Committee on the "1,000 Ship Navy"—A Distributed and Global Maritime Network.

M. Brian Blake is an associate professor and chair of the Department of Computer Science at Georgetown University. With expertise in information technology and in computer science and engineering, he has research interests that include the investigation of automated approaches to sharing information and software capabilities across organizational boundaries, sometimes referred to as enterprise integration. Previously, at Trident Data Systems (now General Dynamics), he was a senior computer scientist responsible for the design and development of applications for the intelligence community; in addition, Dr. Blake served as a software architect at Lockheed Martin Mission Systems, where he managed the object-oriented design of modules in the reengineering of the Global Positioning System upload infrastructure. In the area of service-oriented architecture, Dr. Blake has served as an expert-level systems architect consultant for organizations such as the Department of Defense, the Department of Justice, and the intelligence community. He is a member of the National Science Foundation's Computer and Information Science and Engineering (CISE) Advisory Board.

Clyde G. Chittister is the chief operating officer of the Software Engineering Institute (SEI) at Carnegie Mellon University. He has nearly 40 years of experience in the software and systems engineering fields, including at SEI. Mr. Chittister has held a wide range of management roles, including founder and program director of SEI's Risk Management and Real-Time Systems Programs. He began his career in the field of real-time process-control systems, working on the design, implementation, and maintenance of automated transportation and building conrol systems. He is a senior member of the Institute of Electrical and Electronics Engineers (IEEE) and serves as vice chair of the IEEE Technical Committee on Software Engineering and vice president of finance for the IEEE Systems Council. Mr. Chittister is president-elect of the IEEE Systems Council and an author of numerous published articles on software acquisition, risk management, terrorism, and information technology.

Anup K. Ghosh is a research professor and chief scientist at the Center for Secure Information Systems (CSIS) in the Volgenau School of Information, Technology, and Engineering at George Mason University. Dr. Ghosh's areas of expertise include software security, operating system security, networking security, and malicious code. In addition, he serves as a principal investigator for a multidisciplinary university research initiative aimed at detecting attacks, corruptions, and failures in enterprise-wide servers and client workstations. Prior to joining George Mason University, Dr. Ghosh was a senior scientist and program manager in the Advanced Technology Office at the Defense Advanced Research Projects Agency (DARPA), where he managed an extensive portfolio of information assurance and information operations programs for the Department of Defense.

Raymond Haller is senior vice president and director of MITRE's Department of Defense (DOD) Command, Control, Communications, and Intelligence (C3I) Federally Funded Research and Development Center (FFRDC), where he is responsible for operations, sponsor relations, and advancing the center's overall strategy in information systems technologies. Previously, Mr. Haller was senior vice president of the Command and Control Center, one of two operating centers in the DOD C3I FFRDC, where he was responsible for integration, partnerships, and transformation of military capabilities, including the identification, initiation, and execution of joint C3I activities. Since joining MITRE in 1977, he has held various positions of increasing responsibility and demonstrated an ability to help the government understand the range of technical possibilities while balancing mission needs with cost and technical feasibility.

Richard J. Ivanetich is Institute Fellow at the Institute for Defense Analyses (IDA), having been appointed to that position in 2003. His expertise spans a number of areas of defense systems, technology, and operations analyses, relating primarily to computer and information systems, command-and-control systems and procedures, modeling and simulation of systems and forces, crisis management, and strategic nuclear forces. His previous positions at IDA include service as the director of the Computer and Software Engineering Division and assistant director of the Systems Evaluation Division. Prior to joining IDA in 1975, Dr. Ivanetich was assistant professor of physics at Harvard University. He has served on numerous scientific boards and advisory committees, such as the Cyber and C2 Panels of the U.S. Strategic Command Strategic Advisory Group and as a member of the DARPA Information Science and Technology Study Group; he is a former member of the Naval Studies Board. In 2003, Dr. Ivanetich was elected a National Associate of the National Academies.

John W. Lindquist is president and chief executive officer of EWA Information and Infrastructure Technologies, Inc., a company providing information assurance and information system security engineering services to the government and commercial sectors. He is also chair of the International Systems Security Engineering Association, a not-for-profit organization for systems security engineering. Mr. Lindquist has served on numerous scientific boards and advisory groups, including as a charter member of the Information Technology Sector Coordinating Council and co-chair of its Plans Working Group. Currently he is a member of the Department of Homeland Security's Critical Infrastructure Protection Advisory Committee, a group responsible for developing and implementing the National Infrastructure Protection Plan and the supporting IT Sector Security Plan.

Mark W. Maier is a systems architect and engineer at the Aerospace Corporation. His research areas include systems architecture, radar signal processes, data compression, microsatellites, and computer networks. At the Aerospace Corporation, Dr. Maier developed and now teaches the corporate certificate program in systems architecting. Previously, he was an assistant professor and then an associate professor of electrical and computer engineering at the University of Alabama in Huntsville (UAH). Dr. Maier's work on microsatellites at UAH led to the licensing of a radiation-tolerant computer systems design. Prior to joining UAH, Dr. Maier was an engineer and manager at the Hughes Aircraft Company, where he pioneered an approach to software-based electronic warfare signal analysis that is now widely deployed in production systems.

Richard W. Mayo, VADM, USN (Ret.), is executive vice president for network and enterprise services at CACI International, Inc. In 2004, he retired from the Navy after 35 years of service, concluding as the first commander of the Naval Network Warfare Command, where he was responsible for implementing and securing Navy networks for enhanced warfighter support. Previously, Admiral Mayo served as the director of the Space, Information Warfare, Command and Control Directorate (N6), and as Commander of the U.S. Naval Forces Korea. In addition, from 1993 to 1995, he served as assistant deputy director of the C4 Systems at the Joint Chiefs of Staff.

Ann K. Miller is the Cynthia Tang Missouri Distinguished Professor of Computer Engineering at the Missouri University of Science and Technology. Her areas of expertise include information assurance, with an emphasis on computer and network security; and computer engineering, with an emphasis on large-scale systems engineering, satellite communications, and real-time software. She previously served as deputy assistant secretary of the Navy for research, development, and acquisition (C4I; electronic warfare; and space); Department of the Navy chief information officer; and director for information technologies for DOD research and engineering. Dr. Miller served as a member of the NRC Committee on the Role of Naval Forces in the Global War on Terror.

Daniel M. Schutzer is executive director of the Financial Services Technology Consortium (FSTC), a consortium of banks, financial service providers, national laboratories, and universities, all aimed at addressing strategic business and technology issues, including security and information assurance for the financial sector. Prior to joining FSTC, he served as a director and senior vice president of Citigroup, with responsibilities ranging from trading to retail banking to security and corporate technology. Dr. Schutzer also served as the technical director of Naval Intelligence and Navy Command, Control, and Communications. He has also worked at Sperry Rand, Bell Laboratories, and IBM. He has authored more than 65 publications and 7 books. Dr. Schutzer is a member of the Banking

Industry Technology Sector (BITS) Advisory Council and is a fellow of the New York Academy of Sciences. He served as a member of the NRC Committee on Critical Information Infrastructure Protection and the Law.

Ralph D. Semmel is head of the Applied Information Sciences Department and Infocentric Operations Business Area at the Johns Hopkins University Applied Physics Laboratory (JHU/APL). His areas of expertise include database systems, artificial intelligence, and systems engineering. He previously served at JHU/APL as deputy director of the Research and Technology Development Center and as business area executive for infocentric operations and science and technology, where he established and guided strategic initiatives in global information networks, intelligence systems, information operations, and information assurance. Dr. Semmel also serves as chair of both the computer science and information assurance professional graduates programs at the Johns Hopkins University.

Robert M. Shea, LtGen, USMC (Ret.), is a strategic adviser at Smartronix, a networking and systems management company providing support to military and commercial operations. In 2007, he retired from the U.S. Marine Corps after 36 years of service, concluding as director of C4 Systems at the Joint Chiefs of Staff, where he was the principal adviser to the chairman of the Joint Chiefs of Staff on all C4 matters in the DOD. Previously, General Shea served as the deputy commander for U.S. Forces, Japan; other command positions that he held included commander of the Marine Component to the Joint Task Force Computer Network Defense, director of the Marine Corps Command and Control Systems School, and commanding officer of the 9th Communications Battalion, I Marine Expeditionary Force, during Desert Shield and Desert Storm.

John P. Stenbit (NAE) is an independent consultant whose expertise includes system architectures for complex military and communication systems and systems engineering of information systems. Mr. Stenbit formerly served as assistant secretary of defense for networks and information integration and as the chief information officer for the DOD. Prior to serving in the DOD, he served as executive vice president at TRW, Inc. Mr. Stenbit has served on numerous scientific boards and advisory committees, including as a member of the NRC Committee on Advancing Software-Intensive Systems Productivity. He is a member of the National Academy of Engineering and is a current member of the Naval Studies Board.

Salvatore J. Stolfo is professor of computer science at Columbia University. He received his PhD from New York University Courant Institute in 1979 and has been on the faculty of Columbia ever since. He has published scientific papers in the areas of parallel computing, artificial intelligence knowledge-based systems, data mining, computer security and intrusion, and anomaly detection systems. His most recent research has been in distributed data mining systems with applications to fraud and intrusion detection in network information systems. He has patents in the areas of parallel computing and database inference, Internet privacy, intrusion detection, and computer security. Dr. Stolfo served as chair of the Computer Science Department and director of the Center for Advanced Technology at Columbia University. He recently co-chaired several workshops in data mining, intrusion detection, and the digital government. He is a board member and treasurer of a private organization of Professionals for Cyber Defense. Recently, he participated in a DARPA Innovative Space Based Radar Antenna Technology study and served as an adviser to the director of the DARPA Information Processing Techniques Office as a member of the DARPA Futures Panel.

Edward B. Talbot has been manager of the Computer and Network Security Department at the Sandia National Laboratories (SNL), Livermore, California, since 2006. His areas of expertise include network security operations (wired and wireless) and network architectural needs. Mr. Talbot's responsibilities include managing the Center for Cyber Defenders program. Previously, he was manager of the Advanced Systems Department of the California Weapons Systems Engineering Center, where he developed weapons system concepts and implemented strategies for nuclear deterrence. While at SNL, Mr. Talbot has also worked to develop and implement nuclear systems safety and security enhancements for the current and future nuclear weapons stockpile.

David A. Whelan (NAE) is vice president and deputy general manager, Advanced Systems, and chief scientist, Integrated Defense Systems, at the Boeing Company. His areas of expertise include defense research and development of navigation and timing systems, autonomous air vehicles, and space-based-moving-target indicator radar systems. Prior to joining Boeing, he served as director of the Tactical Technology Office at DARPA. His high-technology development experience includes roles as program manager for the Radar Systems Group of Hughes Aircraft Company, research physicist for the Lawrence Livermore National Laboratory, and a lead low-observables design engineer for B-2A at Northrop Grumman Corporation. Dr. Whelan has served on numerous scientific boards and advisory committees, including the Defense Science Board, the Air Force Scientific Advisory Board, and the NRC Committee on Research, Development, and Acquisition Options for the Special Operations Command (SOCOM). Dr. Whelan is a member of the HRL Laboratories board of directors and the National Academy of Engineering and is currently serving as the vice chair of the Naval Studies Board.

Appendix D

Summary of Recent Naval Operations and Department of Defense Reports Related to Information Assurance

The Committee on Information Assurance for Network-Centric Naval Forces was provided an overview briefing on a number of information assurance studies conducted for the Department of the Navy in recent years.[1] Below is a summary of the most recent revelant reports.[2]

REPORTS PUBLISHED IN 2007

Overview of Data in NCDOC's Prometheus Database

Authors: C.A. Davis and B. Behrens
Abstract: This document catalogs the data that the Navy Cyber Defense Operations Command (NCDOC) currently collects for use in intrusion detection and forensic analysis. The report provides background material for future reference. It documents the source of the data and how they are collected, processed, and ultimately stored in the NCDOC "Prometheus" database.

Operationalizing Information Assurance into Computer Network Defense

Authors: S.W. Young and C.A. Davis
Abstract: The Department of Defense defines the computer network defense

[1]During the course of its study, the committee received (and discussed) materials that are exempt from release under 5 U.S.C. 552(b).
[2]Adapted from information provided to the committee by Michael McBeth, Office of Naval Research Science Advisor, Naval Network Warfare Command, April 28, 2008, Norfolk, Va.

(CND) mission as "actions taken to protect, monitor, analyze, detect, and respond to unauthorized activity within DOD information systems and computer networks." In support of this mission, the Naval Network Warfare Command (NETWARCOM) has drafted a CND concept of operations (CONOPS). The CONOPS lays out a six-step process for CND. As the Navy's CND service provider, the Navy Cyber Defense Operations Command (NCDOC) implements the CND process on Navy-owned networks through its own operational processes and supporting technologies.

Security Information Management for Enclave Networks

Author: R. McQuaid
Abstract: The Air Force enterprise contains networks that are bandwidth-limited, intermittently attached, and/or internally constrained enclaves. These constrained network environments will not support commercial security information management (SIM) feeds and sensors. Recent threat activities have highlighted the need for an information assurance solution that provides consistent SIM-centric monitoring for these enclave networks. This research will improve current SIM deployments within the Air Force by addressing limitations in commercial products. It will influence commercial SIM vendors and the Air Force SIM strategy. By providing IA monitoring to networks that cannot benefit from a centralized SIM, this research will extend the power of SIM technology to the edge of the Air Force enterprise.

Malware Phylogenetics

Authors: P. Chase and D. Beck
Abstract: The nature of malware threats has evolved from widespread outbreaks for the sake of notoriety to large numbers of targeted attacks motivated by economic gain. In this environment it is critical for end users, researchers, investigators, and security tool vendors to have a better understanding of the relationships between malware families and variants in order to improve detection, protection, and response. Understanding the evolutionary relationships between malware threats may provide improved prediction and protection for end users. It may suggest attribution leads and facilitate the reuse of previous analyses by malware analysts and criminal investigators. It could provide a more rigorous basis for naming malware by security vendors, thereby reducing confusion during malware outbreaks and promoting correlation across security tools.

Cross-Boundary Information Sharing

Author: L. Notargiacomo
Abstract: The CIIS Cross Boundary Information Sharing (XBIS) Initiative is a coordinated set of activities at the MITRE Corporation to address critical infor-

mation-sharing problems facing the intelligence community, the Department of Defense, and other MITRE sponsors. This initiative currently focuses on developing an integrated technical laboratory that defines and implements key scenarios that illustrate enablers for and impediments to effective information sharing. The XBIS Laboratory integrates different technologies that enhance information sharing across organizational and classification security boundaries. To demonstrate the capabilities of these technologies, the laboratory provides the ability to simulate many domains and to share information among them. The laboratory architecture supports both integrated scenarios and stand-alone demonstrations, and allows the facility to showcase solutions available today and in the near future.

Navy/OSD Collaborative Review of Acquisition Policy for DoD C3I and Weapons Programs

Authors: D. Gonzales, E. Landree, J. Hollywood, S. Berner, and C. Wong
Abstract: This briefing reviews current U.S. Department of Defense (DOD) policy for ensuring interoperability and information assurance of command, control, communication, intelligence (C3I) and weapons systems. DOD interoperability, information assurance, acquisition, and joint requirement policy are reviewed. This review identifies ambiguities, conflicts, overlaps, and shortfalls in DOD policy and recommends solutions for clarifying policy and remedying other shortcomings. The authors find that interoperability-related policy issuance has sharply increased in recent years and that it includes conflicts and redundancies. They also find that Global Information Grid (GIG) technical guidance is still evolving because of continuing advances and change in networking and software technologies. The authors recommend reducing the number of policies and increasing their actionability and traceability. They also recommend that technology risk levels be developed for GIG functional areas, that these be used to track GIG programs during development, and that network-centric implementation documents more carefully define the capabilities for core GIG enterprise services and specify the technical standards with which GIG programs will have to comply for interoperability.

REPORTS PUBLISHED IN 2006

Alarm Types and Sensor Placement: Effects on Computer Network Defense Operations

Author: S.W. Young
Abstract: In the near future, real-time computer network defense (CND) will be an integral part of military operations. Because the Navy is relying more and more on information technology to move large amounts of data quickly, it must protect that information from compromise, especially when confronting near-peer competitors

with known information operations capabilities. To maintain the confidentiality of plans and operations, the Navy needs a real-time intrusion-detection capability to prevent ongoing attacks from exfiltrating sensitive information such as plans and logistics or denying the use of critical information assets. Today, however, most CND in the Navy is on a non-real-time basis.

A Guide for Assessing Navy Enterprise Information Technology

Authors: J.C. Fauntleroy, L.H. Beard, D.A. Birchler, and L.L. Harle

Abstract: Increasingly, within the vision of network-centric warfare, enterprise networks and capabilities are key to the Navy's achievement of greater coordination and efficiencies in warfare and business functions. To achieve these information technology (IT) and network-related capabilities and efficiencies, expanding enterprise IT (EIT) capabilities must serve the greater needs of the Navy. They must be affordable, given the Navy's many other funding concerns, and adaptable, given the rapid development of new technologies and the many uses for them. The evaluation and assessment of IT and EIT are particularly challenging because of the well-known difficulty in properly estimating return on investment, which lies in the functional mission lanes. From an EIT assessment perspective, there is a lack of visibility into those lanes. The challenge and responsibility to assess EIT investments in the Navy lie with the Assistant Chief of Naval Operations, Information Technology (ACNO-IT), a relatively new organization established to better manage EIT assets and their development. Much of what constitutes EIT in the Navy still resides within the domain of functional area managers, but with the establishment of the ACNO-IT the Navy is seeing a shift in responsibility for enterprise-wide capabilities and their resourcing.

Detecting Malicious Insiders in Military Networks

Author: M. Maybury

Abstract: Given that a network is only as strong as its weakest link, a key vulnerability to network-centric warfare is the threat from within. This paper summarizes several recent efforts of the MITRE Corporation focused on characterizing and automatically detecting malicious insiders (MIs) within modern information systems. Malicious insiders adversely impact an organization's mission through a range of actions that compromise information confidentiality, integrity, and/or availability. Their strong organizational knowledge, varying range of abusive behaviors, and ability to exploit legitimate access make their detection particularly challenging. Crucial balances must be struck while performing MI detection. Detection accuracy must be weighed against minimizing time to detect, and aggregating diverse audit data must be balanced against the need to protect the data from abuse. Key lessons learned from MITRE's MI research include the need to understand the context of the user's actions, the need to establish models of normal behavior, the need to reduce the time to detect malicious behavior, the value of non-cyber-observables, and the importance of real-world data collections to evaluate potential solutions.

Using Honeyclients for Detection and Response Against New Attacks

Author: K. Wang
Abstract: Exploits targeting vulnerabilities in client-side applications are a growing threat on today's Internet. Commonly deployed detection technologies such as honeypots and intrusion-detection systems are useful for detecting server-side attacks but are not effective at detecting client-side attacks. At present there is no proactive client-side attack detection technology. Those using honeyclient technology will gain the capability to proactively detect client exploits in the wild. This project will develop a baseline honeyclient capability and document the ongoing costs of running a honeyclient installation so that informed decisions can be made about how best to apply honeyclient technologies as part of security awareness strategies.

Graph-Based Worm Detection on Operational Enterprise Networks

Authors: D. Ellis, J. Aiken, A. McLeod, D. Keppler, and P. Amman
Abstract: The most significant open challenge to the worm defense community is to develop a sensitive detection method that can detect new worms in real time with a tolerable false-alarm rate. This paper presents a graph-based detection system and validates it on operational enterprise network data. The authors argue that the result is significantly closer to solving this challenge than other published works.

The authors show that a graph-based approach to worm detection in an enterprise network can detect a broad range of active worms with a false-alarm rate of less than two times per day. The supporting analysis comes from running the detection algorithm on a real enterprise network. The sensitivity results are significantly better than what is reported in the literature. The authors can detect all active, fast-spreading unimodal worms, including hit-list, topological, subnet-scanning, and meta-server worms.

REPORTS PUBLISHED IN 2005

Information Technology (IT) Defense, Exploitation, and Attack Study: Identifying Key Maritime IT Domain Technologies for Information Warfare

Author: S.C. Karppi and H. Elitzur
Abstract: At the request of the Office of the Chief of Naval Operations N702, the Center for Naval Analyses (CNA) conducted a study to identify key potential

future U.S. Navy and adversary sea-based/littoral information technologies that, if exploited or attacked, could appreciably alter the Navy's ability to accomplish its Sea Power 21 (SP-21) missions in certain scenarios of interest. The authors refer to those consequential U.S. Navy and adversary technologies as the maritime information technology domain for information operations (IO). Those technologies are ones for which the Navy should build and maintain IO expertise to effectively carry out its SP-21 missions.

Toward More Meaningful Metrics for Computer Network Defense

Authors: D.P. Shea and S.W. Young
Abstract: Developing and implementing a set of practical and informative metrics for computer network defense (CND) pose significant challenges. A computer network, with the associated servers, routers, intrusion detection systems (IDSs), firewalls, and so on generates volumes of data on a daily basis, much of which might be used to form metrics. Likewise, the results of red-team assessments and exercises, and surveys of compliance with Department of Defense CND policies provide additional inputs. The challenges are deciding what decisions can be informed by metrics, selecting the set of variables to track, deciding how to collect and process the data, and finally interpreting the metric outputs and converting these into actionable steps that can head off a network attack or close a security technology gap.

Threats to the GIG and Some Initial Thoughts on Network Security

Authors: A. Hjelmfelt and A.R. Baldwin
Abstract: This document reviews potential threats against Navy information systems, current reports on computer and network incidents, and the types of information assurance practices needed to lessen the risks.

Navy Investments in Computer Network Defense: The Essential Components

Author: S.W. Young
Abstract: The Office of the Chief of Naval Operations N71 asked the Center for Naval Analyses (CNA) to help support the development of an investment strategy for computer network defense (CND). CND is one component of the Information Systems Security Program (ISSP), which is managed by Program Executive Office for Command, Control, Communications, Computers and Intelligence & Space/PMW 160IA and resource sponsored by OPNAV N71. This annotated brief presents some top-level recommendations for technology investments and the associated training programs and policy needed to support a comprehensive CND strategy. In examining technologies, the author uses both the effectiveness and the maturity level of the technologies as a gauge to determine which ones

will be successful at performing the intended mission. Here, "maturity" refers to the experience level of the security community at large in understanding and applying the emerging technologies. "Effectiveness" is assessed by how well the technologies perform their designed tasks. One of the author's fundamental assumptions in performing this analysis is that Internet Protocol version 6 (IPv6) and Internet Protocol Security (IPSec) will be implemented by the Department of Defense as currently planned. The rollout is scheduled to begin in fiscal year 2008. The briefer's recommendations for security technologies are in line with these evolving capabilities.

REPORT PUBLISHED IN 2004

Engaging the Board: Corporate Governance and Information Assurance

Authors: A. Anhal, S. Daman, K. O'Brien, and A. Rathmell
Abstract: This report, prepared for and funded by the Information Assurance Advisory Council, analyzes the relationship between corporate governance and information assurance. The study examines the ways in which information assurance can be embedded into corporate risk management processes in the changing corporate governance environment. Corporate governance now calls for the effective management of risks, but board-level awareness is not yet being translated into effective controls. This study outlines the ways in which information assurance can be embedded into corporate risk management practices and how companies can be incentivized to adopt good practices.

REPORT PUBLISHED IN 2003

The Vulnerability and Assessment Mitigation Methodology

Authors: P. Anton, R. Anderson, R. Mesic, and M. Scheiern
Abstract: Understanding an organization's reliance on information systems and how to mitigate the vulnerabilities of these systems can be an intimidating challenge—especially when considering less-well-known weaknesses or even unknown vulnerabilities that have not yet been exploited. Understanding the risks posed by new kinds of information security threats, the authors build on previous RAND mitigation techniques by introducing the Vulnerability Assessment and Mitigation (VAM) methodology. The six-step procedure uses a top-down approach to protect against future threats and system failures while mitigating current and past threats and weaknesses. The authors lead evaluators through the procedure of classifying vulnerabilities in their systems' physical, cyber, human/social, and infrastructure elements, and of identifying which security techniques can be relevant for these vulnerabilities. The authors also use VAM to break down information compromises into five fundamental components of attack or failure:

knowledge, access, target vulnerability, nonretribution, and assessment. In addition, a new automated tool implemented as an Excel spreadsheet is discussed; this tool greatly simplifies using the methodology and emphasizes analysis on cautions, risks, and barriers.

Appendix E

Naval Information Assurance Architectural Considerations

THE NEED FOR A NAVAL OBJECTIVE ENTERPRISE ARCHITECTURE

As articulated in the 2008 Maritime Strategy,[1] naval forces will provide regionally concentrated, credible combat power as well as globally distributed mission-tailored maritime forces. Some of these missions embody long-standing roles of the Navy, such as power projection and deterrence, while others are a product of globalization and the U.S. role in the world such as humanitarian assistance and disaster response. Some new missions, such as Maritime Domain Awareness (MDA), require the coordination of disparate government organizations, international partners, and industry.[2]

Supporting these goals is complex owing to the diversity and dynamic nature of the missions. Command, control, communications, computers, intelligence, surveillance, and reconnaissance capabilities are important for these missions, and network-centric capabilities are sought in which all nodes contribute to the information superiority of the force. This concept is part of FORCEnet, which is "the operational construct and architectural framework for naval warfare in

[1]ADM Gary Roughead, USN, Chief of Naval Operations; Gen James T. Conway, USMC, Commandant of the Marine Corps; and ADM Thad W. Allen, Commandant of the Coast Guard. 2007. *A Cooperative Strategy for 21st Century Seapower* [Maritime Strategy], Washington, D.C., October. Available at <http://www.navy.mil/maritime/MaritimeStrategy.pdf>. Accessed April 30, 2009.

[2]Honorable Donald C. Winter, Secretary of the Navy. 2008. 2008 Posture Statement of the Honorable Donald C. Winter, Secretary of the Navy, Washington, D.C., February 28. Available at <http://www.navy.mil/navydata/people/secnav/winter/2008_posture_statement2.pdf>. Accessed April 30, 2009.

the Information Age, integrating warriors, sensors, command and control, platforms, and weapons into a networks, distributed combat force."[3] The networking capabilities that FORCEnet must be able to support include a large range of new bandwidth-intensive applications in addition to providing strong protection against adversarial action. Unfortunately, the average age of a typical shipboard network is approaching 7 years, with some as old as 12 years, while industry is operating on a 4-year cycle.[4] The Navy also has several major milestones ahead, such as the introduction of a service-oriented architecture (Consolidated Afloat Networks and Enterprise Services [CANES]), open-architecture computing infrastructure,[5] the follow-on to the Navy/Marine Corps Intranet, NMCI (Next Generation Enterprise Network [NGEN]), and new satellite communications capabilities, all of which require deliberate development and service acquisition strategies owing to their high cost and criticality to naval operations.

Enterprise architecture (EA) provides the discipline for managing change and complexity within the naval enterprise, especially with constrained budgets. EA is required for a truly agile Navy, because the architectural artifacts will allow decision makers to proceed quickly, knowing the state of their information technology (IT) and how it is evolving. Without a fully accepted and leveraged enterprise architecture, agile actions in one part of the enterprise could be detrimental to another part. Another extremely important use of enterprise architecture is in bridging the gap between mission operations and IT implementation. Just as the blueprints for the structure of a house should capture and reflect how the owner desires to operate in the house, the architecture artifacts should be driven by concepts of operations (CONOPS) for the users, should capture and reflect their required capabilities, and should portray back to the owner (funding source) what needs to be built (IT capabilities) to satisfy the mission. Without the EA "bridge," money may be spent on redundant or extraneous IT capabilities.

Enterprise architecture is the practice of applying a comprehensive and rigorous method for describing a current and/or future structure and behavior for an organization's processes, information systems, personnel, and organizational subunits so that they align with the organization's core goals and strategic direction. Although often associated strictly with information technology, EA relates

[3]Admiral Vern Clark, USN, Chief of Naval Operations, and General Michael W. Hagee, USMC, Commandant of the Marine Corps. 2005. "FORCEnet: A Functional Concept for the 21st Century," Department of the Navy, Washington, D.C., February. Available at <http://www.navy.mil/navydata/policy/forcenet/forcenet21.pdf>. Accessed April 30, 2009.

[4]CDR Philip Turner, USN, PMW-160.5, Assistant CANES [Consolidated Afloat Networks and Enterprise Services] Program Manager. 2007. "The CANES Initiative: Bringing the Navy Warfighter onto the Global Information Grid," *CHIPS*, October-December.

[5]For additional discussion, see Open Architecture Enterprise Team, Program Executive Office Integrated Warfare Systems, 2008, The Fourth Quarterly Report to Congress on Naval Open Architecture (NOA), Department of the Navy, Washington, D.C., November.

more broadly to the practice of mission optimization in that it addresses business mission architecture, performance management, and process architecture as well.

An EA is a structured plan for evolving an enterprise from its current state to a desired end state based on enterprise strategies, required capabilities, guiding principles, and external influences. Taking this definition one term at a time:

- The structure is a logically organized rubric or architecture decomposition that clearly shows where all of the components of the enterprise fit within the EA.
- The plan is a set of blueprints for the enterprise along with a transition roadmap or acquisition schedule. It is also beneficial if along with the detailed blueprints there is a corresponding set of high-level artifacts (floor plans and elevations in the building industry) based on the detailed blueprints that aid leadership in decision making.
- The current state is the "as is" architecture or the baseline configuration from which point one needs to migrate. Without this starting point of the journey, the roadmap cannot be drawn.
- The desired end state is the "to be," or target, architecture.
- The enterprise strategies are the mission, vision, goals, and objectives of the enterprise, usually established by the enterprise leadership.
- The required capabilities are the shortfalls or gaps of the as-is architecture that are identified and prioritized by the operations part of the enterprise.
- The guiding principles are tenets of the enterprise that will be used to drive architectural trade-offs and decisions.
- The external influences are architectural drivers such as standards, technology evolution, and the environment within which the enterprise operates.

CURRENT STATE OF NAVY INFORMATION ASSURANCE ARCHITECTURE DEVELOPMENT

The naval forces network can best be described with multiple tiers, a core network, and various types of distribution and access tiers, sometimes called edge networks. The core exists in fiber-optic connectivity, which allows operations in the continental United States as well as at specific fixed regional sites such as military bases. The ability to reach naval forces responding to global conflicts requires communications capabilities beyond core networks, and for this the naval forces rely extensively on satellite communications (SATCOM), with an emphasis on *protected* (extremely high frequency), for assured availability of low-rate information today, to be followed by megabit-class assured connectivity to be provided by the Transformational Satellite Communication System when it is deployed. This capacity is augmented by wideband (super high frequency), narrowband (ultra high frequency), and commercial SATCOM; however, these systems are easily jammed by relatively simple equipment and therefore cannot be relied on for assured connectivity. The Navy's approach to SATCOM is not stovepiped. It includes the integrating element Advanced Digital Network System (ADNS), which adds the essential networking functions on top of the SATCOM links.[6]

A similar degree of diversity is found in terrestrial wireless communications in support of tactical and strategic communications. These missions reflect the operational environments such as Navy battle groups operating in blue water and vessels engaged in littoral operations, as well as United States Marine Corps (USMC) amphibious and ground forces. The Navy also has a critical role in nuclear forces, where architectures must support low-rate but extremely high integrity message transfer. Navy strategic communications to ballistic missile submarines is normally accomplished through a network of very low frequency and low frequency transmitters located throughout the world.

The service, application, and computing elements of the naval architecture are also complex and dynamic (e.g., NMCI migrating to NGEN). The Naval Open Architecture Initiative was established by the Department of the Navy (DON) to shift focus from a platform-centered warfare system acquisition and development approach to an integrated approach centered on the battle force[7,8] In addition, the naval forces have focused on service-oriented architectures as a critical open-architecture technology trend.[9] Several key challenges are associated with the development and deployment of the naval implementation of a service-oriented architecture (SOA), not the least of which is knowing when SOA is the right approach to naval business and weapons system information exchange. Some weapons and combat systems may have latency and data-processing volume requirements that necessitate tightly coupled, real-time, distributed applications and computing components, features that can be difficult to implement based on SOA design principles. However, the potential rigor of SOA configuration control can be useful in such conditions to ensure information availability and integrity; also, the capability of processing is increasing quickly enough that the inefficiencies of SOA could be overcome.

---

[6]National Research Council. 2005. *Navy's Needs in Space for Providing Future Capabilities,* The National Academies Press, Washington, D.C., pp. 216-217.

[7]Naval Surface Warfare Center, Dahlgren Division. 2004. *Open Architecture (OA) Computing Environment Design Guidance, Version 1.0,* Naval Surface Warfare Center (Dahlgren Division), August 23. Available at <http://www.nwsc.navy.mil/TIE/OACE/docs/OACE_Design_Guidance_v1dot0_final.pdf>. Accessed April 30, 2009.

[8]Naval Surface Warfare Center, Dahlgren Division. 2004. *Open Architecture (OA) Computing Environment Design Guidance, Version 1.0,* Naval Surface Warfare Center (Dahlgren Division), August 23. Available at <http://www.nwsc.navy.mil/TIE/OACE/docs/OACE_Design_Guidance_v1dot0_final.pdf>. Accessed April 30, 2009.

[9]Program Executive Office for Integrated Warfare Systems and Open Architecture Enterprise Team. 2007. *Emerging Trends Affecting Future Naval Acquisitions,* Version 7, Washington, D.C., February.

The CANES initiative is a follow-on to the IT-21 Initiative established nearly 10 years ago. The overarching goal of the N6-directed CANES initiative, developed in collaboration with the elements of the Naval NETWAR FORCEnet Enterprise (NNFE), is the same as that of IT-21, which is to establish essential components of the naval afloat (including Maritime Headquarters and Maritime Operations Centers) IT infrastructure that enable deployment of flexible, agile, and cost-effective C4ISR systems and applications. Naval organizations and programs associated with specific domains (i.e., air, space, subsurface, surface, and C4I) have initiated efforts to explore the feasibility of factoring in and deploying key warfighting applications on an SOA.[10]

The naval forces are to be commended for their current enterprise vision[11] and their existing detailed architectures at program levels. They must now ensure that their objective detailed, end-to-end architecture is modified to include the attributes necessary to assure information availability and integrity, and then that they are refined, communicated, accepted broadly, and implemented according to plan. Moreover, a focus on integration and transition from the current state to the desired end state must be maintained.

From an IA perspective, the enterprise architecture must enable naval forces to keep pace with new trends in Department of Defense (DOD) computing, such as the movement toward Web services, and incorporate information assurance mechanisms that deal with threats related to these trends. "Bolting on" IA to naval systems during and after development instead of "building in" IA from system inception appears to be a continuing issue, given the state of implementation relative to the objective enterprise architecture. The problem is exacerbated by quick-reaction development of capabilities and urgent requests from current theaters of operation.

Having overall guidance embodied in an IA set of principles for the naval enterprise architecture could allow IA developers to create solutions that can be more easily and rapidly integrated into naval systems. The prudent, targeted, and timely incorporation of related IA technologies in affected naval IA architectures is therefore an important priority.

In this context, Table E.1 briefly considers selected emerging IA technologies as they relate to each major state of data: in transit, at rest, and in process. An additional category covers technologies and issues that cut across these areas. The naval forces should also seek to leverage testing and evaluation capabilities being developed more broadly by the DOD (e.g., the Defense Advanced Research Projects Agency's National Cyber Range) to evaluate the robustness of architectural implementations and new capabilities, technologies, and systems.

---

[10]"Emerging Trends Affecting Future Naval Acquisitions," Program Executive Office for Integrated Warfare Systems, 7.0, and the Open Architecture Enterprise Team, February 2007.

[11]Victor Ecarma. 2009. "DON Enterprise Architecture Development Supports Naval Transformation," *CHIPS*, Vol. 27, No. 1, January-March, pp. 30-32. Available at <http://www.chips.navy.mil/archives/09_Jan/PDF/enterprise_architecture.pdf>. Accessed April 30, 2009.

TABLE E.1 Selected Emerging Information Assurance Technologies As They Relate to Major States of Data

Technology	Discussion
Relating to Data in Transit	
HAIPE	The evolving High Assurance Internet Protocol Encryptor (HAIPE)[a] standard helps protect data as they transit potentially untrusted networks. HAIPEs are National Security Agency (NSA)-approved Type 1 "in-line" network encryption devices that protect traffic to and from individual devices or entire enclaves. Architects can leverage HAIPE for domain-per-tunnel encryption atop a shared common network backbone. Issues to track include Internet Protocol version 6 compatibility, release of HAIPE 3.0, and the future possibility of HAIPE in software hosted by trusted platforms.
OTNK	Over-the-Network Keying (OTNK) is an approach for establishing cryptographic keys via network-based negotiation rather than by "out-of-band" techniques (e.g., human couriers). Commercial information assurance (IA) protocols (e.g., IPSec, SSL,[b] XMKS[c]) have long employed OTNK techniques, but OTNK for Type 1 keys is an emerging area in Department of Defense (DOD) computing. The coupling of Key Management Infrastructures with OTNK in HAIPE is an important development to monitor.
WS-Security; WS-Policy	Web Service (WS)-Security is a World Wide Web Consortium specification for protecting Web services messages that employ Extensible Markup Language (XML) signature and XML encryption (see below); WS-Policy expresses security requirements between Web services.
Remote attestation	The Trusted Computing Group (TCG) is developing standards (e.g., Trusted Network Connect protocol) related to remote attestation. Remote attestation protocols help parties verify the integrity of remote hosts prior to engagement. The area of attestation depends in turn on trust in the reported results; technologies such as Trusted Platform Modules are useful in establishing trust.
Application firewalls	Firewalls are a first line of defense that block internal assets from external access. The appearance of application-level firewalls to complement their network brethren help secure key communications pathways (e.g., port 80) that are often left open to permit Web traffic. Web services firewalls[d] are an important subcategory of such firewalls.
Relating to Data at Rest	
Domain encryption	Just as specifications such as HAIPE allow domain-specific encryption via network connections, a similar capability is required for data at rest. A number of vendors are working with NSA to develop approved technologies in this area.

continued

TABLE E.1 Continued

Technology	Discussion
Threshold schemes	Threshold schemes are cryptographic approaches for providing both data confidentiality and availability.[e] When coupled with domain encryption, they can provide a level of robustness attractive particularly to tactical environments where connectivity and survivability are key.
Relating to Data in Process	
Virtualization	Virtualization has been in use since the 1960s; however, the recent IA interest in virtualization pertains to allowing multiple separate domains to inhabit the same physical platform. Trust in such virtualization is in part based on trust in the underlying hardware.
TPMs, TXT	An important trend to monitor is the increasing availability of hardware-based security primitives found on mass-produced platforms. Technologies such as TCG's Trusted Platform Modules and Intel's Trusted Execution Technology (TXT) are emerging.
HBSS	The DOD relies extensively on commercial operating systems that have been subject to intensive attacks over many years. To help counter attacks, the Defense Information Systems Agency has selected the McAfee e-Policy Orchestrator suite of tools as part of its Host Based Security Services (HBSS) program[f] to provide a range of protection. HBSS eventually will be deployed throughout the DOD, and, as a result, the Navy will need to keep its systems compatible with the HBSS suite.
RAdAC	Access control is an IA area evolving from relatively static approaches, such as access control lists, to more powerful and general-rule-based approaches. Risk Adaptive Access Control (RAdAC) has been created from a desire to make access control more dynamic.[g]
Web services	XML-based Web services are seen as an important trend in architecting distributed systems that cuts across many areas of IT and IA.[h] Examples of prominent Web services security standards include XML encryption and XML Signature for XML data-level security, the Security Assertions Markup Language (SAML) for expressing security assertions (e.g., assertions for authentication events, attributes, and access decisions), and the Extensible Access Control Markup Language (XACML) for distributed access control.
Policy	Certification and accreditation (C&A) comprise an ever-present issue when new IA technologies are incorporated into DOD IT systems. A key concern has been the cost and time associated with repeated evaluations of such systems, either because of an evolution of the technology used or the context for system use changed. The National Institute of Standards and Technology is currently leading an effort to streamline and standardize the evaluation process across DOD and the intelligence community.[i] In addition, today's high-level policies will have to be revisited and potentially revised as new technologies mature.

TABLE E.1 Continued

Technology	Discussion
Accountability	DOD IT systems have become so complex, interwoven, and dependent on commercial off-the-shelf technologies that they have exceeded their collective ability to confidently assert a guarantee of protection. An open research problem is the development of techniques for measuring assurance that scale to today's complex systems. Lacking such techniques, the secure collection of forensic evidence, including security-critical events, is crucial to help investigators carry out after-the-fact damage assessments. A key challenge, however, is effectively processing a large volume of events. Audit-log reduction tools can help in this regard; they can also help verify that vital services are in use and functioning correctly.

[a]Committee on National Security Systems (CNSS). 2007. *National Policy Governing the Use of High Assurance Internet Protocol Encryptor (HAIPE) Products*, CNSS Policy No. 19, National Security Agency, Ft. Meade, Md., February.

[b]*The Transport Layer Security (TLS) Protocol Version 1.1*, 2006, April. Available at <http://www.ietf.org/rfc/rfc4346.txt>. Accessed August 22, 2008.

[c]World Wide Web Consortium (W3C). 2005. *XML Key Management Specification (XKMS 2.0)*, June 28, 2005. (W3C comprises Massachusetts Institute of Technology, United States; ERCIM [European Research Consortium for Informatics and Mathematics] consisting of 20 countries; and Keio University, Japan). Available at <http://www.w3.org/TR/2005/REC-xkms2-20050628>. Accessed August 22, 2008.

[d]Karen Scarfone and Paul Hoffman, Computer Security Division. 2008. *Guidelines on Firewalls and Firewall Policy*, Special Publication 800-41, Revision 1 (Draft), Information Technology Laboratory, National Institute of Standards and Technology, Gaithersburg, Md., July. Available at <http://csrc.nist.gov/publications/drafts/800-41-Rev1/Draft-SP800-41rev1.pdf>. Accessed April 30, 2009.

[e]Alfred J. Menezes, Paul C. van Oorschot, and Scott A. Vanstone. 1996 (1st ed.), 2001 (5th ed.). *Handbook of Applied Cryptography*, Section 12.7.2., CRC Press, New York.

[f]Defense Information Systems Agency. 2009. "Host Based Security System (HBSS) Fact Sheet," Department of Defense, Washington, D.C. Available at <http://www.disa.mil/news/pressresources/factsheets/hbss.html>. Accessed August 22, 2008.

[g]Gary Machon, National Information Assurance Research Laboratory (NIARL). 2007. "A Mechanism for Risk Adaptive Access Control (RAdAC)," National Security Agency, Ft. Meade, Md., March 14. Available at <www.nsa.gov/SeLinux/papers/radac07.pdf>. Accessed August 22, 2008.

[h]Anoop Singhal, Theodore Winograd, and Karen Scarfone. 2007. *Guide to Web Services Security*, NIST Special Publication 800-95, Computer Security Division, Information Technology Laboratory, National Institute of Standards and Technology, Gaithersburg, Md., August. Available at <http://csrc.nist.gov/publications/nistpubs/800-95/SP800-95.pdf>. Accessed August 22, 2008.

[i]Eustace King. 2008. "Transforming IA Certification and Accreditation Across the National Security Community," *Crosstalk: The Journal of Defense Software Engineering*, July. Available at <http://www.stsc.hill.af.mil/crosstalk/2008/07/0807King.html>. Accessed August 22, 2008.

The naval forces are to be commended for their current enterprise vision[12] and their existing detailed architectures at program levels. They must now ensure that their objective detailed, end-to-end architecture is refined and communicated and accepted broadly. Moreover, a focus on integration and transition from the current state to the desired end state must be maintained.

[12] Victor Ecarma. 2009. "DON Enterprise Architecture Development Supports Naval Transformation," *CHIPS*, Vol. 27, No. 1, January-March, pp. 30-32. Available at <http://www.chips.navy.mil/archives/09_Jan/PDF/enterprise_architecture.pdf>. Accessed April 30, 2009.

Appendix F

Suggested Elements of a Naval Information Assurance Research and Development Program

NETWORK LEVEL

The core fabric of the Internet and the Global Information Grid (GIG) is composed of standard protocols that are vulnerable to exploitation. Sophisticated adversaries, skilled in the art of cyber exploitation and cyberattack, can design their exploits to be difficult to detect. Developing and maintaining survivable networks require secure network functions (routing, addressing) to prevent attacks and to assure correct and attested routing and addressing, as well as countermeasures to defend against successful attacks. Examples of ongoing research that the Navy can build on in this area include the following:

- *BGP/DNS protocol "hardening."* Border Gateway Protocol (BGP) and Domain Name System (DNS) are core network protocols responsible for routing and naming services for all Internet Protocol traffic. Although these protocols have been established and in use for many years at the core of the Internet, a persistent set of vulnerabilities that affect them have been established by the research community with broad and rapid debate about fixes and upgrades. Many experts agree that these core protocols are currently not secure, which means that they can be exploited to reroute traffic to unauthorized destinations in a manner that is not detectable.[1] A number of ongoing research projects from the Department of Homeland Security (DHS) and prior research from the Defense Advanced Research Projects Agency (DARPA) have developed secure implementations of

[1]Joel Hruska. 2008. "Gaping Hole Opened in Internet's Trust-based BGP Protocol," *Ars Technica*, August 27. Available at <http://arstechnica.com/security/news/2008/08/inherent-security-flaw-poses-risk-to-internet-users.ars>. Accessed January 22, 2010.

BGP and DNS, but these have not been adequately vetted and are not broadly deployed. The Office of Management and Budget recently mandated federal adoption of secure DNS.[2] The Navy should be a leader in adopting secure DNS.

- *Network filtering.* Current network filtering strategies tend to be rule-based or signature-specific. A number of research projects at DARPA and the National Science Foundation (NSF) have developed content-based and connection-oriented anomaly detection to detect incoming attacks as well as outgoing exfiltration of sensitive information. Figure F.1 provides a view of one such approach to protecting Web services from cross-site scripting attacks. High-speed networks and encrypted channels complicate matters by exacerbating the problem of content inspection. Consequently, network filtering may have a limited future, forcing the use of technologies that operate closer to the distributed computing nodes at the ends of the network.
- *Network visualization.* Current tools for alerting network operators to attack conditions are text-oriented and voluminous, making the job of understanding the state of the network arduous and error-prone. Network visualization tools exploit a person's capability to process visual cues rapidly for pattern recognition and anomaly detection. Prior and ongoing work at DARPA has developed network visualization tools that can be leveraged to improve the capabilities of network operation centers to detect and respond to attacks.
- *Resilient networks.* In the category of protection, resilient networks ensure that networks can continue to provide service even while under severe denial-of-service attacks. Prior work at DARPA and NSF in overlay networks provides intelligent network elements to detect denial-of-service attacks and automatically throttle traffic to critically needed services.
- *Source attribution.* One of the fundamental limitations of the Internet is that connections are essentially anonymous. The core design of the Internet established a simple means where disparate, geographically and logically separated networks simply announce themselves to one another, and each establishes its own independent routing infrastructure. As a result, it is difficult to ascertain where a connection or an attack is actually coming from, especially when the authority managing a particular network is unfriendly. Source attribution continues to be a continuing research area that the Intelligence Advanced Research Projects Activity (IARPA) is funding.

---

[2]Executive Office of the President, Office of Management and Budget memo, Washington, D.C. August 22, 2008, to Federal Chief Information Officers, requires the adoption of Domain Name System security standards as set forth in National Institute of Standards and Technology (NIST) Special Publication 800-53r1, and that these requirements be fully met by December 2009. See Ron Ross, Stu Katzke, Arnold Johnson, Marianne Swanson, Gary Stoneburner, and George Rogers. 2006. *Recommended Security Controls for Federal Information Systems*, Special Publication 800-53, Revision 1, Computer Security Division, Information Technology Laboratory, National Institute of Standards and Technology, Gaithersburg, Md., December. Available at <http://csrc.nist.gov/publications/nistpubs/800-53-Rev1/800-53-rev1-final-clean-sz.pdf>. Accessed April 30, 2009.

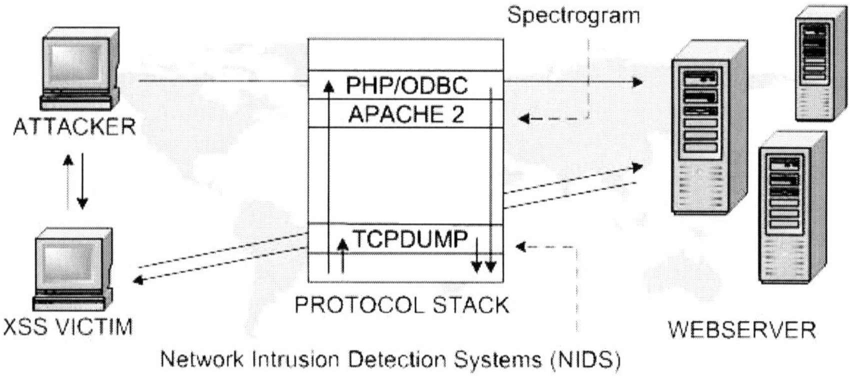

FIGURE F.1 An example Web-layer content sensor and filter. NOTE: Acronyms are defined in Appendix A.

- *Decoy networking.* Sophisticated adversaries will often conduct cyber-based reconnaissance prior to actually attacking. Presenting decoy networks can be an effective strategy for luring an adversary to a fishbowl network isolated from genuine naval forces networks, from which the adversary can be monitored for methods, behavior, and sources. Furthermore, decoy networking may provide a view to an adversary of an arbitrarily large network of bogus but realistic elements that confound and confuse the enemy's attack strategies and targeting. Very little research has been conducted in this area except for work in the area of honeynets and honeypots. Some recent work has been funded partially by DHS and the Army Research Office (ARO). Figure F.2 provides a view of an experimental broadcast decoy injection framework for a wireless fidelity (WIFI) network.

SYSTEM LEVEL

Information technology (IT) systems composed of many distributed components, perhaps each with varying levels of security, pose serious information assurance (IA) problems. Large collections of common components provide a severe threat from a single common attack that may lead to catastrophic consequences, but also an opportunity that may also be leveraged to enhance security. Research topics in this area include the following:

- *Secure composition.* Today, a single vulnerable software component in a system can compromise the integrity of an entire system. Research in the secure composition of distributed components, funded by NSF, aims to enable the composition of components into systems in which security properties of the whole are guaranteed, or at least bounded. Such means are assumed to have been solved in

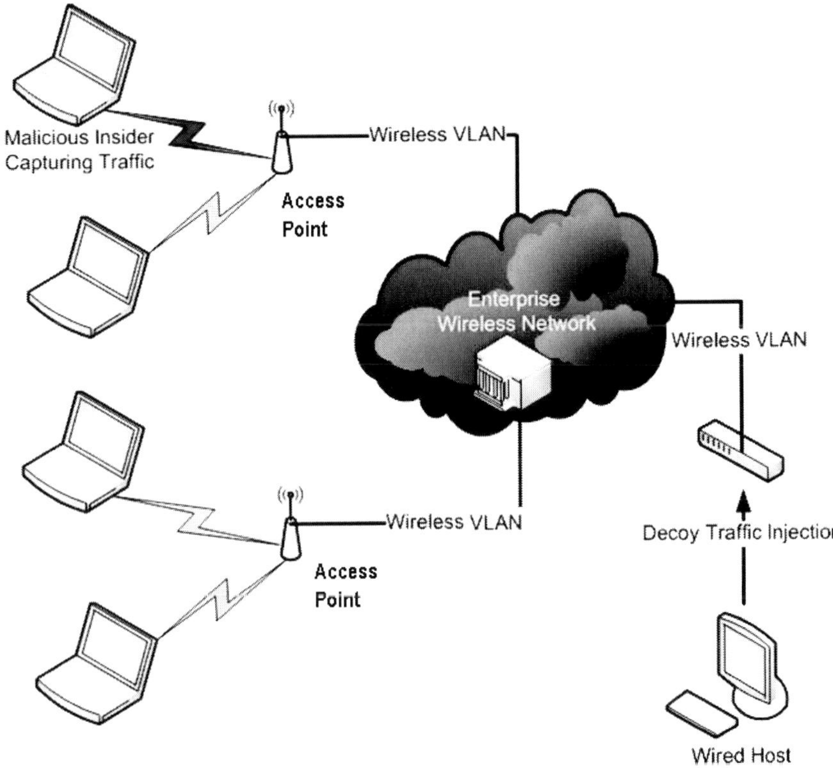

FIGURE F.2 A decoy- or bait-injection framework. NOTE: Acronyms are defined in Appendix A.

the long-term vision of the GIG in the context where deep application knowledge may be required for effective composition. The problem is far more difficult than simply defining a set of interface policies.

- *Artificial diversity.* Military and federal networks as a whole are currently actively managed to be uniformly homogeneous. This makes them easier to manage on the one hand, but on the other, uniformly susceptible to a single contagion. To break monoculture and increase resiliency, artificial diversity techniques funded by DARPA introduce diversity into the computing fabric; these techniques permit applications to interoperate, but change the structural properties of code to make different instances of the same software diverse in implementation.
- *Collaborative software communities.* While monocultures pose a risk as described above, some DARPA-funded work in application communities and related research funded by NSF have turned this vulnerability into a potential IA asset. This is accomplished by making each instance of the common software a

sensor on the network, dynamically sharing attack data with other instances in order to responsively harden other instances of the software against in-progress attacks that they may also experience. Research focused on developing a number of related security-alert-sharing technologies (that maintain privacy across administrative domains) have also been sponsored by NSF and DHS.

- *Privacy-preserving technologies.* Security of systems requires confidentiality of data. Encryption logically serves as a fundamental capability, but it is insufficient, especially in the context of applications in which data are shared across domains with various levels of mutual (dis-)trust. This notion is extended to query processing, whereby questions posed by an organization that seeks data about some topic may also be considered as confidential. IARPA at present sponsors work in secure multiparty computation and privacy-preserving technologies permitting enclaves to share data securely and privately without revealing what information is sought by either party. These technologies promise to allow effective sharing while maintaining strict compartmentalization.

HOST LEVEL

The fundamental IA challenge remains at the end points of networks. The core host software platforms and applications present a constant flow of discovered vulnerabilities that can be exploited by a persistent adversary in possession of the necessary skills and resources. A generation ago the technical principles of object-oriented programming were developed, whereby systems can be dynamically composed of objects that permit the reuse of software and the sharing of passive and active data among software components. Embedded in the design capabilities afforded by object-oriented design methods is the ability to dynamically communicate, interpret, and execute software among distributed computing components—that is, modern object-oriented systems provide code injection platforms. Injected code may be benign and useful (such as JavaScript drawing a table of information on a Webpage), or malicious and harmful (such as a Trojan embedded in a host by a malicious e-mail attachment). Furthermore, driven by customer demand and time-to-market considerations, commercial application vendors typically introduce products to market that are less than sufficiently tested, evaluated, and debugged, thereby providing sophisticated adversaries with the opportunity to exploit software design flaws that have not been discovered by the vendor prior to product release.

Much of the response by the commercial security marketplace has been to provide signature-based detection and filter solutions requiring the continual updating of a growing signature base for known software exploitations. The inevitable response by sophisticated adversaries is to generate new attack vectors for which no signatures are yet available. This cat-and-mouse game was quite manageable, since the time from discovering a vulnerability to the time of generating an attack vector to exploit that vulnerability was measured in time frames of weeks to days. New attack tools have clearly shifted the balance to the attacker in two ways. First, design patterns for attack tools have been developed to allow the rapid creation of zero-day attack vectors; second, tools have been designed to allow the generation of a very large set of variants that can avoid discovery, thereby forcing a defense that would need to look for an unmanageable number of attack signatures. In summary, signature-based defenses will become technically obsolete, while current IA architecture designs are dependent on such defenses.

Furthermore, the offshore outsourcing of development, both hardware and software, exacerbates the problem by providing ample opportunity for a sophisticated adversary purposely to embed its attack vectors into commercial off-the-shelf (COTS) products that are regularly procured by the Department of Defense (DOD). To counter this fundamental danger of commercial IT practice, a number of advanced concepts to harden the host and improve the security of its software are being actively pursued. Topics include methods to create new secure and safe software and to automate security policy implementation. Many methods have been proposed to create secure software, but these do not adequately address the huge legacy-software base that now runs and operates modern enterprise systems, and the Internet in use today. A few representative research topics that deal with improving the security of systems broadly in use are enumerated below:

- *Counter-evasion techniques for obfuscated malware.* Given the obsolescence of signature-based technologies, new and effective methods to identify malware embedded in content flows are required to keep pace with the advances made by sophisticated adversaries. Rich content flows, including Web pages, documents, and other media, may legitimately include code for transfer to a recipient computer. Automatically determining the intent of code remains an open research problem, to distinguish malice from useful function. Furthermore, adversaries have cleverly obfuscated and embedded malicious code in content streams where code is not ordinarily expected. Detecting these stealth-attack vectors remains an open research problem.
- *Virtualization for security.* Virtualization technology has been widely adopted for server consolidation and is beginning to be adopted to support multilevel security needs. However, virtualization can also be used to isolate untrusted applications from the host operating system. For example, an application can be considered to be untrusted if it communicates to untrusted networks (such as the Non-Classified Internet Protocol Router Network), runs untrusted content (such as media files from an untrusted source), or has unknown provenance. DARPA-funded work has developed application-level virtualization that seamlessly virtualizes applications transparently to users to isolate untrusted applications from trusted systems and networks.
- *Self-healing software.* Substantial progress has been made in designing software that monitors and models its own behavior. This line of work on anomaly detection has been extended recently by work funded by DARPA and the Air Force

Office of Scientific Research (AFOSR) to develop techniques so that software is self-aware of its own operation in order to detect violations of its integrity and repair itself after attack, leaving it more robust after attack, similar to human immune systems.

- *Hardware life-cycle tamper resistance.* DARPA's Trust in Integrated Circuits program is developing techniques to detect compromises in chip-level designs and implementations during supply chain life-cycle attacks. Far more of an investment is needed in this line of work to develop tamper-resistant hardware designs.

USER LEVEL

Many IA research and development (R&D) researchers have come to agree that system users constitute a core security threat, primarily owing to error and mistakes, but also to purposeful malfeasance. The insider attack threat has been known for quite some time but has not been adequately addressed. A growing body of literature is now appearing that recognizes this vexing security problem. Considerable R&D is needed in this area, including the following:

- *Behavior-based security.* One of the most effective techniques for detecting insider threats is to analyze user behavior patterns for inappropriate access of network resources such as file servers, printers, and outbound connections. Ongoing work at the MITRE Corporation employs Bayesian analysis of user behavior to detect certain insider threats with a reasonably high reliability. Far more research is needed in order to understand user intent for detecting malicious or dangerous actions. Limited work is being sponsored by DHS and ARO in this area.
- *Defense through uncertainty.* An emerging area, initially funded by IARPA and AFOSR, this topic leverages uncertainty in deployed environments to make it difficult for an adversary to exploit them. Knowledge and information about the target environment are sufficiently "fuzzed," confusing the attacker to confound the intended end goals. One example is to present purposely erroneous server operating system images for entities connected on the network. This can result in an intended attack being delivered to an incorrect operating system environment. Another example is using decoy documents placed intelligently in a network so that if the documents are exfiltrated, the home organization will be aware of the theft but the adversary will not realize their false pretense. Many other opportunities to confound and confuse an enemy are possible leveraging the principle of uncertainty. Of course, the use of these tactics requires management and control processes to ensure that desired activities are not inadvertently disrupted.

PRIVILEGED USER LEVEL

Perhaps the most vexing and difficult security problem is best captured by the adage "Who checks the checkers?" Security personnel are extremely privileged users with access to all key functions of the enterprise system. A recent example of the malfeasance in this area involved a system administrator who captured San Francisco's entire administrative IT infrastructure and denied access to all system administrators but himself.[3] Critical weapons systems are designed with safety systems and technologies that inhibit a single insider from unauthorized action, but little work has been done in the research community to address the core question of how to secure security systems from security and operating personnel who are the deepest insiders and who potentially pose the insider threat with the highest risk.

- *Role- and behavior-based access control.* A fundamental tenet of IA is that data and applications are only accessed by authenticated and authorized users who require access to conduct their business. The pervasive use of access controls based on credentials (IDs, passwords, and pins) is woefully inadequate in complex network environments. Role-based access control considers means of associating the logical roles of a user with the specific data and applications used by the specific roles defined with an enterprise. Research in this area by NSF has been extended by DARPA and some industrial laboratories also to associate "behavior" with a user's credentials as a means of granting access to network resources.
- *Self-protecting security technologies.* In much the same way that networks are threatened by denial-of-service attacks, host-based security technologies are threatened by denial-of-sensor attacks. A user may disable a host security system by accident, or a system administrator may bypass a security subsystem by design. This threat is just beginning to be recognized in the research community, and some work is proposed that deals with security technologies that are protected from this threat. Work done at the Sandia National Laboratories on safety technologies for nuclear weaponry may be brought to bear on this underfunded area of research related to the insider threat.

[3]Ashley Surdin. 2008. "San Francisco Case Shows Vulnerability of Data Networks," *Washington Post,* August 11, p. A03. Available at <http://www.washingtonpost.com/wp-dyn/content/article/2008/08/10/AR2008081001802.html>. Accessed March 16, 2009.